T0211582

Lecture Notes in Artificial Intelligence 10459

Subseries of Lecture Notes in Computer Science

More information about this series at http://www.springer.com/series/1244

Andrey Ronzhin · Gerhard Rigoll
Roman Meshcheryakov (Eds.)

Interactive Collaborative Robotics

Second International Conference, ICR 2017
Hatfield, UK, September 12–16, 2017
Proceedings

 Springer

Editors
Andrey Ronzhin (iD)
SPIIRAS
St. Petersburg
Russia

Roman Meshcheryakov (iD)
Tomsk State University
Tomsk
Russia

Gerhard Rigoll (iD)
Technical University of Munich
Munich
Germany

ISSN 0302-9743 ISSN 1611-3349 (electronic)
Lecture Notes in Artificial Intelligence
ISBN 978-3-319-66470-5 ISBN 978-3-319-66471-2 (eBook)
DOI 10.1007/978-3-319-66471-2

Library of Congress Control Number: 2017949516

LNCS Sublibrary: SL7 – Artificial Intelligence

Printed on acid-free paper

This Springer imprint is published by Springer Nature
The registered company is Springer International Publishing AG
The registered company address is: Gewerbestrasse 11, 6330 Cham, Switzerland

Preface

The International Conference on Interactive Collaborative Robotics (ICR) was established as a satellite event of the International Conference on Speech and Computer (SPECOM) by St. Petersburg Institute for Informatics and Automation of the Russian Academy of Science (SPIIRAS, St. Petersburg, Russia), the Technical University of Munich (TUM, Munich, Germany), and Tomsk State University of Control Systems and Radioelectronics (TUSUR, Tomsk, Russia) in 2016.

The conference gathered together experts from different communities to discuss challenges facing human and robot collaboration in industrial, social, medical, educational, and other areas. Models of collaborative behavior of one or multiple robots with physical interaction with people in operational environment equipped embedded sensor networks and cloud services in conditions of uncertainty and environmental variability are the main issues of the conference.

ICR 2017 was the second event in the series and this time it was organized by the University of Hertfordshire, in cooperation with SPIIRAS, TUM, and TUSUR. The conference was held during September 12–16, 2017 at the College Lane campus of the University of Hertfordshire in Hatfield, UK, 20 miles (30 kilometers) north of London, just 20 minutes by train from London's King's Cross station.

This volume contains a collection of submitted papers presented at the conference, which were thoroughly reviewed by members of the Program Committee consisting of around 20 top specialists in the conference topic areas. Theoretical and more general contributions were presented in common (plenary) sessions. Problem oriented sessions as well as panel discussions then brought together specialists in limited problem areas with the aim of exchanging knowledge and skills resulting from research projects of all kinds.

Last but not least, we would like to express our gratitude to the authors for providing their papers on time, to the members of the conference reviewing team and Program Committee for their careful reviews and paper selection, and to the editors for their hard work preparing this volume. Special thanks are due to the members of the Local Organizing Committee for their tireless effort and enthusiasm during the conference organization. We hope that you benefitted from the event and that you also enjoyed the social program prepared by the members of the Organizing Committee.

September 2017

Andrey Ronzhin
Gerhard Rigoll
Roman Meshcheryakov

Organization

The conference ICR 2017 was organized by the University of Hertfordshire, in cooperation with St. Petersburg Institute for Informatics and Automation of the Russian Academy of Science (SPIIRAS, St. Petersburg, Russia), Munich University of Technology (TUM, Munich, Germany), and Tomsk State University of Control Systems and Radioelectronics (TUSUR, Tomsk, Russia). The conference website is located at: http://specom.nw.ru/icr.

Program Committee

Roman Meshcheryakov, Russia
  (Co-chair)
Gerhard Rigoll, Germany (Co-chair)
Andrey Ronzhin, Russia (Co-chair)
Christos Antonopoulos, Greece
Sara Chaychian, UK
Dimitrios Kalles, Greece
Igor Kalyaev, Russia
Alexey Kashevnik, Russia
Gerhard Kraetzschmar, Germany

Dongheui Lee, Germany
Iosif Mporas, UK
Vladmir Pavlovkiy, Russia
Viacheslav Pshikhopov, Russia
Yulia Sandamirskaya, Switzerland
Jesus Savage, Mexico
Hooman Samani, Taiwan
Evgeny Shandarov, Russia
Lev Stankevich, Russia

Organizing Committee

Iosif Mporas (Chair)
Polina Emeleva
Alexey Karpov
Dana Kovach

Ekaterina Miroshnikova
Andrey Ronzhin
Dmitry Ryumin
Anton Saveliev

Contents

Applying MAPP Algorithm for Cooperative Path Finding in Urban Environments

Anton Andreychuk[1,2(✉)] and Konstantin Yakovlev[1,3]

[1] Federal Research Center "Computer Science and Control" of Russian
Academy of Sciences, Moscow, Russia
yakovlev@isa.ru, andreychuk@mail.com
[2] People's Friendship University of Russia (RUDN University), Moscow, Russia
[3] National Research University Higher School of Economics, Moscow, Russia

Abstract. The paper considers the problem of planning a set of non-conflict trajectories for the coalition of intelligent agents (mobile robots). Two divergent approaches, e.g. centralized and decentralized, are surveyed and analyzed. Decentralized planner – MAPP is described and applied to the task of finding trajectories for dozens UAVs performing nap-of-the-earth flight in urban environments. Results of the experimental studies provide an opportunity to claim that MAPP is a highly efficient planner for solving considered types of tasks.

Keywords: Path planning · Path finding · Heuristic search · Multi-agent path planning · Multi-agent path finding · MAPP

1 Introduction

Planning a set of non-conflict trajectories for a coalition of intelligent agents is a well known problem with the applications in logistics, delivery, transport systems etc. Unlike path-finding for a single agent (robot, unmanned vehicle etc.) for which numerous computationally-effective algorithms exists [1–4] the multi-agent path finding problem lacks an efficient general solution and belongs to the PSPACE hard problems [5]. Therefore the development of computationally efficient algorithms for cooperative path finding is actual yet challenging task.

Commonly in robotics and Artificial Intelligence path finding is considered to be a task of finding the shortest path on a graph which models agent's environment. Vertices of such graph correspond to the locations of the workspace the agent can occupy and edges correspond to the transitions between them, e.g. to the elementary trajectories the agent can follow in an automated fashion (segments of straight lines, curves of the predefined shape etc.). In 2D path finding, which is under consideration in this work, the most often used graph models are: visibility graphs [6], Voronoi diagrams [7], navigation meshes [8], regular grids [9]. The latter are particularly widespread in robotics [10] as they can be constructed (and updated) on-the-fly when processing the sensors input.

© Springer International Publishing AG 2017
A. Ronzhin et al. (Eds.): ICR 2017, LNAI 10459, pp. 1–10, 2017.
DOI: 10.1007/978-3-319-66471-2_1

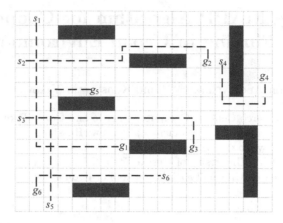

Fig. 1. Graphical representation of the grid based cooperative path finding task. Light cells correspond to traversable vertices, dark cells - to untraversable ones. Dotted lines show agents' paths

Formally, grid is a weighted undirected graph. Each element (vertex) of the grid corresponds to some area of regular shape (square in considered case) on the plane, which can be either traversable or untraversable for the agent (mobile robot, UAV etc.). In this work we assume that grid vertices are placed in the centers of the squares the agents workspace is tessellated into. The edges are pairs of cardinally adjacent traversable vertices. Graphically, grid can be depicted as a table composed of black and white cells (see Fig. 1). Path on a grid is a sequence of grid edges between two defined vertices, e.g. *start* and *goal*. When augmenting path with time we end up with the trajectory. Typically, in grid based path finding time is discretized into timesteps 1, 2, 3, ..., and agent is allowed to traverse a single grid edge per one timestep or stand still occupying some vertex (grid cell). We follow this approach in the paper as well and state the cooperative path finding as defined in Sect. 2. We then overview the existing methods to solve the problem (Sect. 3) and describe one of the prominent algorithm, e.g. MAPP, tailored to solve it (Sect. 4). We experimentally evaluate MAPP in solving navigation tasks for the coalitions of the UAVs (up to 100) performing nap-of-the-earth flight in urban environments and provide the results of this evaluation in Sect. 5. Section 6 concludes.

2   Formal Statement

Consider a square grid which is a set of cells, L, composed of two mutually disjoint subsets, e.g. blocked (untraversable) cells, L^-, and unblocked (traversable) cells, $L^+ : L = L^+ \cup L^- = \{l\}$. U – is a non-empty set of agents navigating on the grid. Each agent is allowed to perform cardinal moves, e.g. can go left, right, up or down (from one traversable cell to the other). Path for a single agent u from given start and goal cells, e.g. s_u and g_u, is a sequence of cardinally adjacent traversable cells: $\pi(u) = (l_0^u, l_1^u, \ldots, l_{k_u}^u), l_0^u = s_u, l_{k_u}^u = g_u, \forall l_i^u \in \pi(u) : l_i^u \in L^+, l_i^u \in c_{adj}(l_{i-1}^u)$. Here $c_{adj}(l)$ denotes a set of cardinally adjacent cells for the cell l. At the initial moment of time the agents are located at the corresponding start cells. During a timestep each agent can move from the cell l_i^u to the cell l_{i+1}^u or stay at place. The set of moments of time (timesteps) at which agent u occupies the cell l_i^u is denoted as $time(l_i^u)$. The cooperative path finding task is to build a set of non-conflicting trajectories, i.e. such paths that the following conditions are met: $\forall u, v \in U, \forall l_i^u = l_j^v : time(l_i^u) \cap time(l_j^v) = \emptyset, i \in [0, k_u], j \in [0, k_v]$.

3 Methods and Algorithms

Cooperative (multi-agent) path finding algorithms can be divided into two groups: centralized (coupled) and decentralized (decoupled) planners. Centralized planners search trajectories for all agents simultaneously, taking into account the position of every agent at each moment of time. Such algorithms are theoretically complete, e.g. guarantee finding a solution if it exists, and can be optimal. Decentralized planners typically build on top of assigning priorities to the agents, finding trajectories one-by-one according to these priorities and eliminating conflicts. They are much faster than the centralized ones but do not guarantee optimality (and even completeness) in general.

One of the well known optimal algorithm implementing the centralized approach is the OD + ID algorithm described in [11]. OD + ID builds on top of A* [12]. The distinctive feature of OD + ID is the use of the procedures that allow speeding up the search and reducing the search space. The first procedure, Independence Detection (ID), divides the original task into several subtasks in such a way that the optimality of the sought solution is not violated. The second - Operator Decomposition (OD), is used to reduce the search space. Instead of moving all agents at the same time, agents move in turn, thereby reducing the number of different states from M^n to $M*n$, where M is the number of agents and n is the number of possible moves for each agent (4 in considered case).

One of the most advanced centralized planners up-to-date is the CBS (Conflict Based Search) algorithm [13] and its variations like ICBS [14]. It's a two level search algorithm. At the low level, it searches a path for a single agent and every time it finds a conflict, it creates alternative solutions. At the high level, the algorithm builds a tree of alternative solutions until all conflicts are eliminated. Due to the fact that among the alternative solutions the best is always chosen, the optimality of the overall solution is guaranteed. Two level decomposition leads to reducing the search space and increasing the efficiency of the algorithm as compared to other centralized planners.

Another well-known centralized and optimal planner is M* [15]. It relies on the technique called subdimensional expansion, which is used for dynamical generation of low dimensional search spaces embedded in the full configuration space. It helps to reduce the search space in some cases. But if the agents cannot be divided into independent subgroups (and this can happen quite often in practice), the introduced technique doesn't work well and M* spends too much time to find a solution.

In general, optimal algorithms for cooperative path planning that implement centralized approach are very resource-intensive and do not scale well to large problems when the number of agents is high (dozens or hundreds) and the search space consists of tens and even hundreds of thousands of states.

There exist a group of centralized algorithms that trade off the optimality for computational efficiency.

One of such algorithms is OA (Optimal Anytime) [16], proposed by the authors of OD + ID. The principle of its work is similar to the OD + ID algorithm. The difference is that when conflicts are detected, agents do not try to build joint optimal trajectories. They join the group, but build their own paths, considering other agents only as moving obstacles. This approach works faster, but does not guarantee the optimality of the solution any more.

There also exist suboptimal variants of CBS, such as ECBS [17], for example.

The second large class of the cooperative path finding algorithms is a class of the algorithms utilizing a decentralized approach. These algorithms are based (in one or another way) on splitting the initial problem into several simpler problems which can be viewed as a form of prioritized planning first proposed in [18]. The algorithms implementing a decentralized approach are computationally efficient, but do not guarantee finding the optimal solutions and are not even complete in general.

One of the first and most applicable algorithms in the gaming industry, using a decentralized approach, is the LRA* (Local Repair A*) algorithm proposed in [19]. The algorithm works in two stages. At the first stage, each agent finds an individual path using A* algorithm, ignoring all other agents. At the second stage, agents move along the constructed trajectories until a conflict arises. Whenever an agent cannot go any further, because another agent is blocking its path, the former re-plans its path. How-ever, due to the fact that each agent independently reshapes its path, this leads to the emergence of new conflicts and as a result, there may arise deadlocks (cycle conflicts) that cannot be resolved.

WHCA* (Windowed Hierarchical Cooperative A*) algorithm described in [20] is one of the algorithms eliminating the shortcomings of LRA* algorithm. Like LRA*, WHCA * algorithm works in two stages and uses the A* algorithm to find individual paths. The improvement of the algorithm lies in the fact that the search for the path for each agent occurs in the "window" - an abstract area of the environment model (grid) having the size $w * w$, which is gradually shifts towards the goal. The algorithm uses a 3-dimensional space-time table, in which the positions of all agents are stored at each timestep. Using such a table can reduce the number of conflicts that arise between agents, thereby increasing the speed of the algorithm. However, WHCA* algorithm is also not complete, i.e. in some cases it does not find a solution, even though it exists. Another disadvantage of the algorithm is its parametrization, i.e. the dependence on the parameter w.

One of the last modifications of WHCA* – Conflict Oriented WHCA* (CO-WHCA*) [21] combines ideas from both WHCA* and LRA*. It uses a technique that focuses on areas with potential conflicts between agents. The experiments carried out by the authors of the algorithm showed that it finds more solutions than WHCA* and LRA*, but is still not complete.

There also exist decentralized (decoupled) planners that guarantee completeness under well-defined constraints [22,23]. These algorithms are of particular interest to our study as they scale well to large problems, find solutions fast while the restrictions they impose on class of tasks they guarantee to solve are likely to be held in the scenarios we are looking at (navigating dozens of UAVs in urban environments). We choose MAPP algorithm for further consideration as it is originally tailored to solve multi-agent path finding tasks on 4-connected grid-worlds (as in considered case).

4   Algorithm MAPP

MAPP algorithm is a prominent cooperative path planner [22,24] that utilizes decentralized approach, but in contrast to other decentralized algorithms mentioned above, is complete for a well-defined class of problems called Slidable.

An agent $u \in U$ is Slidable if a path $\pi(u) = (l_0^u, l_1^u, \ldots, l_{k_u}^u)$ exists and the next three conditions are met:

1. Alternative connectivity. For each three consecutive locations $l_{i-1}^u, l_i^u, l_{i+1}^u$ on $\pi(u)$ there exists an alternative path Ω_i^u between l_{i-1}^u and l_{i+1}^u that does not include $l_i^u, i = \overline{(1, k_u - 1)}$ (see Fig. 2).
2. Initial blank. At the initial timestep, when all agents are at l_0^u, l_1^u is blank, i.e. unoccupied, for each agent $u \in U$.
3. Goal isolation. Neither π nor Ω-paths include goal locations of other agents. More formally:
   (a) $\forall v \in U \backslash \{u\} : g_u \notin \pi(v)$
   (b) $\forall v \in U, \forall i = \overline{(1, k_v - 1)} : g_u \notin \Omega_i^v$.

Problem belongs to the class Slidable if all agents $u \in U$ are Slidable.

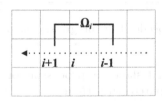

Fig. 2. An example of an alternative path Ω_i

At the first stage MAPP finds individual Slidable paths with the enhanced A* algorithm. The modifications of the original A* [12] are the following. First MAPP's A* checks for a blank location in the first step. Second it always avoids passing through other goals. Third it ensures that alternate connectivity condition is always met, e.g. when expanding a node x', its neighbor x'' is added to the set of potential candidates (so-called OPEN list) only if there exists an alternate path between x'' and x, the parent of x'. Thus, MAPP guarantees that each node in path π (if it is found) can be excluded and replaced by the alternative path Ω (see Fig. 2).

After the first stage, all agents, for which Slidable paths are found, move step by step to their goals in the predefined order (from high-priority agents to low priority ones). To ensure that lower priority agent do not interfere with any of the higher priority agents a concept of the private zone is used. The private zone includes the current location of the agent and the previous position in case an agent is moving along the computed path. Disallowing the lower priority agents to move through the private zones of higher priority agents one can guarantee that at least one agent with the highest priority will move closer to its goal at each step. The procedure for progression step is described in Algorithm 1. The figure is taken from [24] (Fig. 3).

Algorithm 1 Progression step.

1: **while** changes occur **do**
2:    **for** each $u \in A$ in order **do**
3:       **if** $\mathrm{pos}(u) \notin \pi(u)$ **then**
4:          do nothing {u has been pushed off track as a result of blank travel}
5:       **else if** u has already visited l^u_{i+1} in current progression step **then**
6:          do nothing
7:       **else if** the next location, l^u_{i+1}, belongs to the private zone of a higher priority unit, i.e. $\exists v < u : l^u_{i+1} \in \zeta(v)$ **then**
8:          do nothing {wait until l^u_{i+1} is released by v}
9:       **else if** l^u_{i+1} is blank **then**
10:          move u to l^u_{i+1}
11:       **else if** can bring blank to l^u_{i+1} **then**
12:          bring blank to l^u_{i+1}
13:          move u to l^u_{i+1}
14:       **else**
15:          do nothing

Fig. 3. Pseudocode of the progression step (from [24])

Lines 3–15 in Algorithm 1 show the processing of current agent u. If u has been pushed off its precomputed path, then it stays in its current location and does nothing (lines 3–4). Lines 5 and 6 cover the situation when agent u has been pushed around (via blank travel) by higher-priority agents back to a location on $\pi(u)$ already visited in the current progression step. If u is on its path but the next location l^u_{i+1} is currently blocked by an agent v with higher priority, then it just waits (lines 7–8). Otherwise, if the next location l^u_{i+1} is available, u moves there (lines 9–10). Finally, if l^u_{i+1} is occupied by a lower priority agent, then it brings blank to l^u_{i+1} and moves agent u there (lines 11–13). When u moves to

l_{i+1}^u, algorithm checks if l_{i+1}^u is the goal location of u. If this is the case, then u is marked as solved by removing it from A and adding it to S, the set of solved agents.

MAPP algorithms guarantees that all Slidable agents will reach their destinations. The proof of this statement is based on the fact that at least one agent, the one with the highest priority, gets closer to its goal at each step. After a finite number of steps this agent will reach the goal, the highest priority will go to some other agent and so on. Thus at the end all agents will achieve their goals without colliding with other agents. Non-Slidable agents will simply stay at their start locations.

5   Experimental Evaluation

We were interested in scenarios of finding non-conflict trajectories for numerous (dozens of) UAVs performing nap-of-the-earth flight in urban areas. The considered scenario assumes that all UAVs fly at the same altitude and that allows us to consider the problem as a search for non-conflict trajectories on the plane. It was shown earlier in [13] that existing optimal algorithms based on the centralized approach, are poorly applicable for the problems involving large number (>85) of agents acting in similar environments, as they need too much time (more than 5 min) to solve a task. Therefore, the main goal was to examine whether MAPP algorithm will succeed. We fixed the number of agents (UAVs) to be 100.

To run experiments MAPP algorithm was implemented from scratch in C++. The experiments were conducted using Windows-7 PC with Intel QuadCore CPU (@ 2.5 GHz) and 2 GB of RAM.

The algorithm was tested on a collection of 100 urban maps. Each map represents a $1.81\,\mathrm{km}^2$ fragment of the real city environment as it was extracted from the OpenStreetMaps geo-spatial database[1]. An example is depicted on Fig. 4. Each map was discretized to a 501×501 grid. All 100 maps were chosen in such a way that the number of blocked cells was 20–25% on each of them.

Path finding instances (sets of start/goal locations) of two types were generated (type-1 and type-2 tasks). For type-1 instances start and goal locations were chosen randomly at the opposite borders of the map (see Fig. 4-left), resulting in sparse distribution of the start/goals. For the type-2 instances compact start and goal zones on the opposite borders of the map were chosen and all start/goals were confined to these zones (see Fig. 4-right). The size of each zone was 50×50 cells. For each map two instances of each type were generated. Thus the testbed consisted of 400 tasks in total.

To assess the efficiency of the algorithm, the following indicators were tracked:

- Time - algorithm's runtime. Measured in seconds.
- Memory - the maximum number of nodes that the algorithm stores in memory during the path planning for an individual agent.

---

[1] http://wiki.openstreetmap.org/wiki/Database.

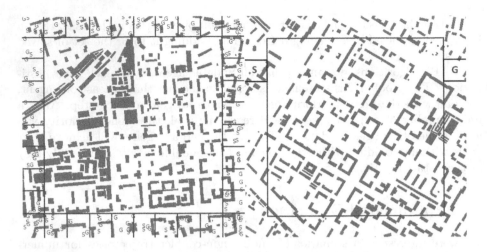

Fig. 4. Fragments of the city maps that were used for experimental evaluation

- Path length - the number of transitions between the grid cells plus the waiting time, averaged for all solved agents of the instance.
- Conflicts - the number of times high-priority agent is blocked by the low-priority one.
- Conflict agents - number of the agents that have at least one conflict.
- Success - the percentage of agents for which Slidable trajectories were successfully found.

Table 1 presents the averaged results of the experiment. First row in the table contains results for the type-1 tasks, second row for the type-2 tasks.

As one can see, MAPP successfully solved all the type-1 instances (when the agents were disseminated sparsely close to the map borders), although for the type-2 tasks the success rate was 99.97%. In some cases (6, to be precise) MAPP was not able to find Slideble trajectory for one agent and this was enough to deem the task failed.

Obviously, type-2 tasks are more complex in terms of conflict resolution. All agents move from one constrained zone to another and therefore their trajectories often overlap. Due to this, there are almost 6 times more conflicts (compared to type-1 instances) and more than 85% of agents have at least one conflict. However, this almost does not affect the runtime, since most of the time is spent on building the individual trajectories.

Table 1. Averaged results of experimental evaluation of MAPP

#	Time (s)	Memory (nodes)	Pathlength	Conflicts	Success
1	32.606	84955	540.457	33.015	100%
2	31.398	59362	540.559	192.31	99.97%

Solving type-2 tasks consumes notably less memory. This can be explained in the following manner. Start and goal locations for type-1 tasks were located in different areas of maps. Thus for almost all tasks, chances are at least one of the agents needs to build complex trajectory, enveloping a large number of obstacles, to accomplish its individual task. Building such complex paths necessarily leads to processing (and storing in memory) a large number of search nodes. On the opposite, individual trajectories for type-2 tasks were approximately the same, due to the fact start and goals for all agents were located in the same compact regions. Therefore, for some tasks, when there were few obstacles between the start and goal zones, the memory consumption was small and this influenced the final averaged value.

Obtained results show that MAPP algorithm can be successfully applied to solving both types of tasks. The performance indicators of the algorithm are high showing its computational efficiency for solving considered class of cooperative path finding scenarios.

6   Conclusion

In this paper the problem of cooperative path finding for a set of homogeneous agents was considered. An analysis of existing methods for solving this problem was presented. MAPP algorithm was described, implemented and evaluated experimentally. The results of the conducted experiments confirmed its ability to efficiently solve cooperative path finding problems for large number (one hundred) of UAVs performing nap-of-the-earth flight in urban environments.

Acknowledgements. This work was supported by the Russian Science Foundation (Project No. 16-11-00048).

References

1. Koenig, S., Likhachev, M.: D* lite. In: Proceedings of the AAAI Conference of Artificial Intelligence (AAAI), pp. 476–483 (2002)
2. Magid, E., Keren, D., Rivlin, E., Yavneh, I.: Spline-based robot navigation. In: Proceedings of 2006 IEEE/RSJ International Conference on Intelligent Robots and Systems (IROS), Beijing, China, 9–15 October 2006, pp. 2296–2301 (2006)
3. Yakovlev, K., Baskin, E., Hramoin, I.: Grid-based angle-constrained path planning. In: Hölldobler, S., Krötzsch, M., Peñaloza, R., Rudolph, S. (eds.) KI 2015. LNCS, vol. 9324, pp. 208–221. Springer, Cham (2015). doi:10.1007/978-3-319-24489-1_16
4. Harabor, D., Grastien, A., Öz, D., Aksakalli, V.: Optimal any-angle pathfinding in practice. J. Artif. Intell. Res. (JAIR) **56**, 89–118 (2016)
5. Hopcroft, J., Schwartz, J., Sharir, M.: On the complexity of motion planning for multiple independent objects; PSPACE-hardness of the "warehouseman's problem". Int. J. Robot. Res. **3**(4), 76–88 (1984)
6. Lozano-Pérez, T., Wesley, M.A.: An algorithm for planning collision-free paths among polyhedral obstacles. Commun. ACM **22**(10), 560–570 (1979)

7. Bhattacharya, P., Gavrilova, M.L.: Roadmap-based path planning - using the Voronoi diagram for a clearance-based shortest path. IEEE Robot. Autom. Mag. **15**(2), 58–66 (2008)
8. Kallmann, M.: Navigation queries from triangular meshes. In: Boulic, R., Chrysanthou, Y., Komura, T. (eds.) MIG 2010. LNCS, vol. 6459, pp. 230–241. Springer, Heidelberg (2010). doi:10.1007/978-3-642-16958-8_22
9. Yap, P.: Grid-based path-finding. In: Cohen, R., Spencer, B. (eds.) AI 2002. LNCS, vol. 2338, pp. 44–55. Springer, Heidelberg (2002). doi:10.1007/3-540-47922-8_4
10. Elfes, A.: Using occupancy grids for mobile robot perception and navigation. Computer **22**(6), 46–57 (1989)
11. Standley, T.: Finding optimal solutions to cooperative pathfinding problems. In: Proceedings of The 24th AAAI Conference on Artificial Intelligence (AAAI-2010), pp. 173–178 (2010)
12. Hart, P., Nilsson, N., Raphael, B.: A formal basis for the heuristic determination of minimum cost paths. IEEE Trans. Syst. Sci. Cybern. **4**(2), 100–107 (1968)
13. Sharon, G., Stern, R., Felner, A., Sturtevant, N.R.: Conflict-based search for optimal multi-agent pathfinding. Artif. Intell. **219**, 40–66 (2015)
14. Boyarski, E., Felner, A., Stern, R., Sharon, F., Tolpin, D., Betzalel, D., Shimony, S.: ICBS: improved conflict-based search algorithm for multi-agent pathfinding. In: Proceedings of the 24th International Joint Conference on Artificial Intelligence (IJCAI-2015), pp. 740–746 (2015)
15. Wagner, G., Choset, H.: M*: a complete multirobot path planning algorithm with performance bounds. In: Proceedings of The 2011 IEEE/RSJ International Conference on Intelligent Robots and Systems (IROS-2011), pp. 3260–3267 (2011)
16. Standley, T., Korf, R.: Complete algorithms for cooperative pathfinding problems. In: Proceedings of The 22d International Joint Conference on Artificial Intelligence (IJCAI-2011), vol. 1, pp. 668–673. AAAI Press (2011)
17. Barer, M., Sharon, G., Stern, R., Felner, A.: Suboptimal variants of the conflict-based search algorithm for the multi-agent pathfinding problem. In: Proceedings of the 7th Annual Symposium on Combinatorial Search (SOCS-2014), pp. 19–27 (2014)
18. Erdmann, M., Lozano-Pérez, T.: On multiple moving objects. Algorithmica **2**, 1419–1424 (1987)
19. Zelinsky, A.: A mobile robot exploration algorithm. IEEE Trans. Robot. Autom. **8**(6), 707–717 (1992)
20. Silver, D.: Cooperative pathfinding. In: Proceedings of the 1st Conference on Artificial Intelligence and Interactive Digital Entertainment (AIIDE-2005), pp. 117–122 (2005)
21. Bnaya, Z., Felner, A.: Conflict-oriented windowed hierarchical cooperative A*. In: Proceedings of the 2014 IEEE International Conference on Robotics and Automation (ICRA-2014), pp. 3743–3748 (2014)
22. Wang, K.-H.C., Botea, A.: Tractable multi-agent path planning on grid maps. In: Proceedings of the 21st International Joint Conference on Artificial Intelligence (IJCAI-2009), pp. 1870–1875 (2009)
23. Čáp, M., Novák, P., Kleiner, A., Selecký, M.: Prioritized planning algorithms for trajectory coordination of multiple mobile robots. IEEE Trans. Autom. Sci. Eng. **12**(3), 835–849 (2015)
24. Wang, K.-H.C., Botea, A.: MAPP: a scalable multi-agent path planning algorithm with tractability and completeness guarantees. J. Artif. Intell. Res. (JAIR) **42**, 55–90 (2011)

Real-Time Removing of Outliers and Noise in 3D Point Clouds Applied in Robotic Applications

Gerasimos Arvanitis[(✉)], Aris S. Lalos, Konstantinos Moustakas, and Nikos Fakotakis

Electrical and Computer Engineering Department, University of Patras, Rio, Patras, Greece
{arvanitis,aris.lalos,moustakas,fakotaki}@upatras.gr

Abstract. Nowadays, robots are able to carry out a complex series of actions, to take decisions, to interact with their environment and generally to perform plausible reactions. Robots' visual ability plays an important role to their behavior, helping them to efficiently manage the received information. In this paper, we present a real time method for removing outliers and noise of 3D point clouds which are captured by the optical system of robots having depth camera at their disposal. Using our method, the final result of the created 3D object is smoothed providing an ideal form for using it in further processing techniques; namely navigation, object recognition and segmentation. In our experiments, we investigate real scenarios where the robot moves while it acquires the point cloud in natural light environment, so that unpleasant noise and outliers become apparent.

Keywords: Robotic vision · 3D point clouds denoising · Outliers removing

1 Introduction

The sense of sight plays an important role to human's life. It allows us to connect with our surroundings, keep us safe and help us to learn more about the world. In the same way, vision is also important for robots, affecting their behavior especially when they try to interact with their environment. Robots have become smarter and their behavior is more realistic approaching humans, in many ways. One basic reason leading to this success is the learning process which is an off-line task, though. In real time applications, robots must discover their world using only their own sensors and tools without an a priori knowledge. Undoubtedly, the type and the precision of sensors affect robot's perception and behavior. In recent years, a lot of research has been carried out into the field of robots vision, having presented excellent results applied to applications like segmentation, recognitions and others. Although there has been significant work in this way, some limitations still remain. Image analysis of the conventional RGB cameras, used mostly at the previous years, does not provide all the necessary information existing in

© Springer International Publishing AG 2017
A. Ronzhin et al. (Eds.): ICR 2017, LNAI 10459, pp. 11–19, 2017.
DOI: 10.1007/978-3-319-66471-2_2

the real world, like the 3D perception. On the other hand, depth cameras have become popular as an accessible tool for capturing information. Nevertheless, depth cameras are very sensitive to motion and light conditions and this may cause abnormalities to the final results, especially when robot moves while it captures the point cloud.

2   Related Works

Depth cameras have started to be used in a lot of robotic applications providing extra useful information (geometric and topographic) which is not able to be provided by the conventional cameras. Robots need the 3D visual perception in order to understand better their environment and demonstrate a more satisfied behavior. A detailed vision capability is necessary in many challenging applications especially when further processing tasks are required; like segmentation, object recognition and navigation. For example, robots need to intelligibly specify and identify their environment (borders, existing objects, obstacles) before they start to navigate inside it.

In [1] a method for registration of 3D point clouds is presented. In order to identify a complete point cloud presentation of the robot gripper it is rotated in front of a stereovision camera and its geometry is captured from different angles. However this approach does not use filters to remove noise from the data. The paper in [2] also presents point cloud registration algorithms which are used in mobile robotics. In this work [3] authors use a hand-held RGB-D camera to create a 3D model. The research in [4] shows the capability of using a mobile robot for performing real-time vision tasks in a cloud computing environment, however without any pre-processing step taking place. In [5], an object recognition engine needs to extract discriminative features from data representing an object and accurately classify the object to be of practical use in robotics. Furthermore, the classification of the object must be rapidly performed in the presence of a voluminous stream of data. The article in [6] investigates the problem of acquiring 3D object maps of indoor household environments.

The analysis of the above state-of-the-art works makes it clear that a different approach should be used for real time applications. If a robot moves, while it receives points, then its line-of-sight focus might become unstable creating outliers to the acquired 3D point cloud. Additionally, different light conditions can affect the final results. The proposed approach manages to handle these challenging situations creating smoothed 3D objects by removing outliers and noise.

The rest of this paper is organized as follows: Sect. 2 presents an overview of our method. Section 3 presents our experimental results. Section 4 draws the conclusions.

3   Overview of Our Method

In this section we describe the main steps followed by the proposed method. We also introduce the used mathematical background and all the necessary assumptions. Firstly, we assume that the robot captures visual information by collecting points using its depth camera. However, in the general case the precision of cameras is limited or a relative motion between robot and target exists, and as a result the acquired 3D point cloud suffers from noise, outliers and incomplete surfaces. These problems must be solved before robot uses this inappropriate information. In the next paragraphs we describe how our method manages to overcome these limitations.

3.1   Framework Overview

In Fig. 1 the basic steps of our approach are briefly presented. Firstly, the robotic optical sensor starts capturing points of objects that exist in robot's line of vision. The acquired 3D point clouds are usually noisy and they also have misaligned outliers. We use Robust PCA for removing the outliers and consequently we create a triangulated model based on k nearest neighbors (k-NN) algorithm. The triangulation process helps us to specify the neighbors of each point so that the bilateral filtering method can efficiently be used as denoising technique. At the end, a smoothed 3D mesh is created which has a suitable form for being used by other applications or processing tasks.

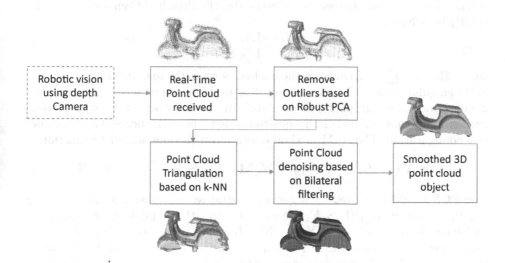

Fig. 1. Framework of our proposed method

3.2   Removing 3D Point Cloud Outliers

Let us assume that the captured 3D point cloud $\mathbf{M}$ consists of m points represented as a vector $\mathbf{v} = [\mathbf{x}, \mathbf{y}, \mathbf{z}]$ in a 3D coordinate space $\mathbf{x}, \mathbf{y}, \mathbf{z} \in \Re^{m \times 1}$ and $\mathbf{v} \in \Re^{m \times 3}$. Some of these points are considered as outliers and they must be removed. The method that we use for the removing of outliers is the robust

PCA [7]. We start by assuming that the point cloud, represented as a matrix of data $\mathbf{M}$, may be decomposed as:

$$\mathbf{M} = \mathbf{L} + \mathbf{S} \tag{1}$$

where $\mathbf{L}$ is a low-rank matrix representing the space of real data while $\mathbf{S}$ is a sparse matrix representing the space where outliers lie. The dimensions of $\mathbf{L}$ and $\mathbf{S}$ matrices as well as the location of their non-zero entries are considered as unknown. The classical Principal Component Analysis (PCA) estimates the $\mathbf{L}$ by solving:

$$\begin{aligned} minimize \quad & \|\mathbf{M} - \mathbf{L}\| \\ subject\ to\ & rank(\mathbf{L}) \leq m \end{aligned} \tag{2}$$

Assuming that there are few outliers and also that the entries of $\mathbf{S}$ have an independent and identically Gaussian distribution then the above equation can be efficiently solved using the singular value decomposition (SVD). Nonetheless, this approach is not appropriate for this case study. The problem of outliers in the point cloud can be considered as an idealized version of Robust PCA. Our purpose is to recover the low-rank matrix $\mathbf{L}$ from the measurements $\mathbf{M} = \mathbf{L} + \mathbf{S}$ which are highly corrupted by the matrix $\mathbf{S}$ with randomly large magnitude of entries. The approach that we use [8] estimates the Principal Component Pursuit (PCP) by solving:

$$\begin{aligned} minimize\ & \|\mathbf{L}\|_* + \lambda\|\mathbf{S}\|_1 \\ subject\ to\ & \mathbf{L} + \mathbf{S} = \mathbf{M} \end{aligned} \tag{3}$$

where $\|\mathbf{L}\|_* := \sum_i \sigma_i(\mathbf{L})$ denotes the nuclear norm of the matrix which is the sum of the singular values of $\mathbf{L}$. This convex PCP problem can be solved using an augmented Lagrange multiplier (ALM) algorithm, as described in [9], which works with stability across a wide range of situations without the necessity of parameters configuration. The ALM method operates on the augmented Lagrangian:

$$l(\mathbf{L}, \mathbf{S}, \mathbf{Y}) = \|\mathbf{L}\|_* + \lambda\|\mathbf{S}\|_1 + <\mathbf{Y}, \mathbf{M} - \mathbf{L} - \mathbf{S}> + \frac{\mu}{2}\|\mathbf{M} - \mathbf{L} - \mathbf{S}\|_F^2 \tag{4}$$

The PCP would be solved by a generic Lagrange multiplier algorithm setting $(\mathbf{L}_i, \mathbf{S}_i) = \arg \min_{\mathbf{L},\mathbf{S}} l(\mathbf{L}, \mathbf{S}, \mathbf{Y}_i)$ repeatedly, and then updating the Lagrange multiplier matrix via $\mathbf{Y}_{i+1} = \mathbf{Y}_i + \mu(\mathbf{M} - \mathbf{L}_i - \mathbf{S}_i)$. This process is applied in m overlapped areas, where the i-th area a_i consists of vertex $\mathbf{v}_i$ and its 30 nearest neighbors vetrices. In our case study, where a low-rank and sparse decomposition problem exists, the $min_{\mathbf{L}} l(\mathbf{L}, \mathbf{S}, \mathbf{Y})$ and $min_{\mathbf{S}} l(\mathbf{L}, \mathbf{S}, \mathbf{Y})$ both have very simple and efficient solutions. We introduce two ancillary operators $\mathcal{Q}_\tau$ and $\mathcal{D}_\tau$. $\mathcal{Q}_\tau : \Re \longrightarrow \Re$ denotes the shrinkage operator $\mathcal{Q}_\tau[.] = sgn(.)max(|.| - \tau, 0)$ and extends it to matrices by applying it to each element while the $\mathcal{D}_\tau(.)$ denotes the singular value thresholding operator given by $\mathcal{D}_\tau(.) = U\mathcal{Q}_\tau(\sum)V^T$. The estimation occurs according to:

$$\arg \min_{\mathbf{S}} l(\mathbf{L}, \mathbf{S}, \mathbf{Y}) = \mathcal{Q}_{\lambda\mu^{-1}}(\mathbf{M} - \mathbf{L} + \mu^{-1}\mathbf{Y}) \tag{5}$$

$$\arg \min_{\mathbf{L}} l(\mathbf{L}, \mathbf{S}, \mathbf{Y}) = \mathcal{D}_{\mu^{-1}}(\mathbf{M} - \mathbf{S} + \mu^{-1}\mathbf{Y}) \tag{6}$$

Firstly, the l is minimized with respect to $\mathbf{L}$ (fixing $\mathbf{S}$) and then the l is minimized with respect to $\mathbf{S}$ (fixing $\mathbf{L}$). Finally, the Lagrange multiplier matrix $\mathbf{Y}$ is updated based on the residual $\mathbf{M} - \mathbf{L} - \mathbf{S}$. The process is repeated for every area a_i $\forall\, i = 1, m$.

$$\mathbf{S} = \mathbf{Y} = 0; \mu > 0$$
$$\mathbf{L}_{i+1} = \mathcal{D}_{\mu^{-1}}(\mathbf{M} - \mathbf{S}_i + \mu^{-1}\mathbf{Y}_i);$$
$$\mathbf{S}_{i+1} = \mathcal{Q}_{\lambda\mu^{-1}}(\mathbf{M} - \mathbf{L}_{i+1} + \mu^{-1}\mathbf{Y}_i); \qquad (7)$$
$$\mathbf{Y}_{i+1} = \mathbf{Y}_i + \mu(\mathbf{M} - \mathbf{L}_{i+1} + \mathbf{S}_{i+1});$$

The cost of each iteration is estimating $\mathbf{L}_{i+1}$ using singular value thresholding. This makes necessary the computation of those singular vectors of $\mathbf{M} - \mathbf{S}_i + \mu^{-1}\mathbf{Y}_i$ whose corresponding singular values exceed the threshold μ^{-1}. There are two important implementation that need to be chosen; namely, the choice of μ and the stopping criterion. The value of μ is chosen as $\mu = m^2/4\|\mathbf{M}\|_1$ [10], while the iterative process terminates when $\|\mathbf{M} - \mathbf{L} - \mathbf{S}\|_F \leq \delta\|\mathbf{M}\|_F$, with $\delta = 10^{-7}$. After this step the vertices decrease duo to the removing of the outliers, so the number of the remaining vertices is m_r where $m_r < m$.

3.3   Triangulation Based on k-NN

The acquired 3D point cloud is unorganized, meaning that the connectivity of its points is unknown thus we are unable to take advantage of the geometric relation between neighboring vertices. To overcome this limitation we estimate the connectivity between points based on k-NN algorithm using $k = 7$ as the length of each cell, as a matter of fact each point is connected with the 7 nearest neighbor points. The defined connectivity helps us to create the binary adjacency matrix $\mathbf{C} \in \Re^{m_r \times m_r}$ consisting of the following elements:

$$c_{ij} = \begin{cases} 1 \; if \; i, j \in \mathbf{r}_i \\ 0 \; otherwise \end{cases} \qquad (8)$$

where $\mathbf{r}_i$ represents the first ring area of vertex i consisting of its k nearest neighboring vertices, so that $\mathbf{r}_i = [r_{i1}, r_{i2}, \cdots, r_{ik}]$. The 1s of this sparse matrix represents edges; by connecting 3 corresponding vertices a face is created such that $f_i = [\mathbf{v}_{i1}, \mathbf{v}_{i2}, \mathbf{v}_{i3}] \; \forall\, i = 1, m_f$ where $m_f > m_r$. We also estimate the normals of each face based on the below formula:

$$\mathbf{n}_{fi} = \frac{(\mathbf{v}_{i_2} - \mathbf{v}_{i_1}) \times (\mathbf{v}_{i_3} - \mathbf{v}_{i_1})}{\|(\mathbf{v}_{i_2} - \mathbf{v}_{i_1}) \times (\mathbf{v}_{i_3} - \mathbf{v}_{i_1})\|} \; \forall\, i = 1, m_f \qquad (9)$$

where $\mathbf{v}_{i_1}, \mathbf{v}_{i_2}, \mathbf{v}_{i_3}$ represents the vertices that are related with the face f_i.

3.4   3D Mesh Denoising Using Bilateral Filtering

The final step of our approach is the mesh denoising using the bilateral filtering algorithm directly applied to the triangulated 3D point cloud. The used approach of the bilateral algorithm, as described in [11,12], is a fast method and it can be easily implemented in real time applications. The denoising process occurs by firstly estimating the d factor according to Eq. 10 and then updating the noisy vertices according to Eq. 11.

$$d_i = \frac{\sum_{j \in k_i} W_p W_n h}{\sum_{j \in k_i} |W_p W_n|} \tag{10}$$

where $W_p = e^{\frac{-t^2}{2\sigma_p^2}}$, $W_n = e^{\frac{-h^2}{2\sigma_n^2}}$, $t = \|\mathbf{v}_i - \mathbf{v}_j\|$, $h = <\mathbf{n}, \mathbf{v}_i - \mathbf{v}_j>$ and $\mathbf{n} = \frac{\sum_{j \in k_i} \mathbf{n}_f}{k}$ $\forall\, i = 1, m_r,\ \forall\, j \in k_i$. Once the $\mathbf{d} = [d_1\ d_2\ \cdots\ d_{m_r}]$ is estimated then each vertex is updated according to:

$$\hat{\mathbf{v}} = \mathbf{v} + \mathbf{n} \cdot \mathbf{d}^T \tag{11}$$

where $\hat{\mathbf{v}} \in \Re^{m_r \times 3}$ represents the vector of smoothed vertices. This method manages to preserve the features of the object without further shrinking of the already smoothed areas.

4   Experimental Analysis and Results

In this section we present the results of our approach following the steps of the proposed method as described above. For the evaluation we used a variety of data in different light and motion conditions. Here we present two corresponding examples using models of two different geometrical categories. The first model consists of flat surfaces (Moto) while the second has a lot of features and details (Cactus), as presented in Fig. 2.

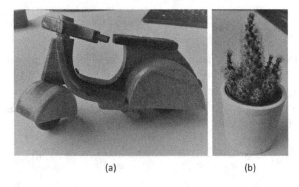

(a)   (b)

Fig. 2. Images of the original objects used as models for the evaluation process, (a) Moto (b) Cactus

In order to establish experimental conditions as close as possible to the real cases, the experiments have taken place using only natural lights, no artificial

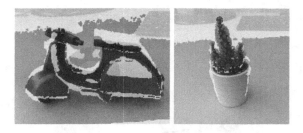

Fig. 3. The moment when the capturing point process starts

lights or other additional light sources are used. Figure 3 depicts the moment when robot's vision starts, while in Fig. 4, snapshots of the capturing process are illustrated for the two models.

Figure 4 depicts two different case studies of acquiring points. In the first case Fig. 4(a), a carefully gathering takes place while in the second case Fig. 4(b), a sudden motion of the robot causes noise and outliers. Nevertheless, observing carefully the figures, we can see that even in the first case there are imperfections that need to be covered.

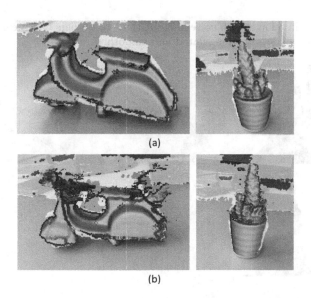

Fig. 4. (a) Stable collection of points, (b) Collection of points with a sudden robot's movement

The step-by-step extracted results of the proposed method are presented in Fig. 5. More specifically, Fig. 5(a) presents the point clouds having noise and outliers, as received by the depth camera sensor. In Fig. 5(b) the outliers of the

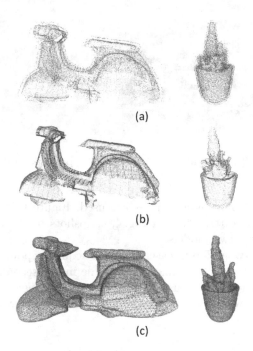

(a)

(b)

(c)

Fig. 5. (a) 3D point clouds with noise and outliers, (b) outliers have been removed but noisy parts still remain, (c) triangulated and smoothed results

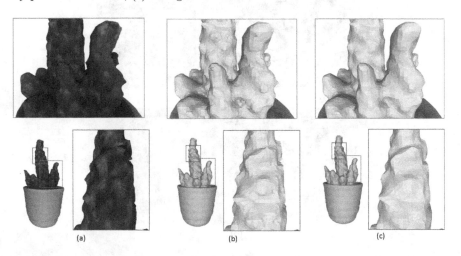

(a)   (b)   (c)

Fig. 6. (a) Cactus model is presented having its original texture, (b) noisy surfaces after outliers' removing and triangulation process, (c) the 3D denoised mesh based on bilateral filtering

point clouds have been removed, however abnormalities of noise still remain. In Fig. 5(c) the final triangulated and smoothed objects are presented.

In Fig. 6, the smoothed results using bilateral filtering for the Cactus model are presented. In addition, enlarged presentation of details are also shown for an easier comparison between the 3D objects.

5   Conclusions

In this paper we presented a method, ideally suited for real-time applications. The proposed approach can be used for removing outliers and noise that lie in 3D point clouds, captured by robot's depth cameras. Additionally, the method handles problems like the relative motion between robot and its capturing target and also the natural light conditions which can cause abnormalities in the acquired point cloud. Using the proposed approach, we manage to overcome these limitations and as a result a smoothed 3D mesh is created which has an ideal form for further processing usage.

References

1. Jerbić, B., Šuligoj, F., Švaco, M., Šekoranja, B.: Robot assisted 3D point cloud object registration. Procedia Eng. **100**, 847–852 (2015). ISSN 1877-7058
2. Pomerleau, F., Colas, F., Siegwart, R.: A review of point cloud registration algorithms for mobile robotics. Found. Trends® Robot. **4**(1), 1–104 (2015)
3. Kidson, R., Stanimirovic, D., Pangercic, D., Beetz, M.: Elaborative evaluation of RGB-D based point cloud registration for personal robots. In: ICRA 2012 Workshop on Semantic Perception and Mapping for Knowledge-Enabled Service Robotics, 14–18 May 2012
4. Beksi, W.J., Papanikolopoulos, N.: Point cloud culling for robot vision tasks under communication constraint. In: 2014 IEEE/RSJ International Conference on Intelligent Robots and Systems, Chicago, IL, pp. 3747–3752 (2014)
5. Beksi, W.J., Spruth, J., Papanikolopoulos, N.: CORE: a cloud-based object recognition engine for robotics. In: 2015 IEEE/RSJ International Conference on Intelligent Robots and Systems (IROS), pp. 4512–4517 (2015)
6. Rusu, R.R., Marton, Z.C., Blodow, N., Dolha, M., Beetz, M.: Towards 3D point cloud based object maps for household environments. Robot. Auton. Syst. **56**(11), 927–941 (2008)
7. Aravkin, A., Becker, S., Cevher, V., Olsen, P.: A variational approach to stable principal component pursuit. In: Zhang, N., Tian, J. (eds.) Proceedings of the Thirtieth Conference on Uncertainty in Artificial Intelligence (UAI 2014), pp. 32–41. AUAI Press, Arlington (2014)
8. Candès, E.J., Li, X., Ma Y., Wright, J.: Robust principal component analysis? J. ACM **58**(3), Article 11 (2011)
9. Lin, Z., Chen, M., Wu, L., Ma, Y.: The augmented Lagrange multiplier method for exact recovery of a corrupted low-rank matrices. In: Mathematical Programming (2009, submitted)
10. Yuan, X., Yang, J.: Sparse and Low-Rank Matrix Decomposition Via Alternating Direction Methods, November 2009. optimization-online.org
11. Zheng, Y., Fu, H., Au, O.K.C., Tai, C.L.: Bilateral normal filtering for mesh denoising. IEEE Trans. Vis. Comput. Graph. **17**(10), 1521–1530 (2011). doi:10.1109/TVCG.2010.264
12. Fleishman, S., Drori, I., Cohen-Or, D.: Bilateral mesh denoising. In: ACM SIGGRAPH 2003 Papers (SIGGRAPH 2003), pp. 950–953. ACM, New York (2003)

Emotion Recognition System
for Human-Robot Interface: Comparison
of Two Approaches

Anatoly Bobe[1(✉)], Dmitry Konyshev[2], and Sergey Vorotnikov[2]

[1] Neurobotics Ltd., 4922 passage, 2/4, Zelenograd,
Moscow 124498, Russian Federation
a.bobe@neurobotics.ru
[2] BMSTU Robototechnika, Izmaylovskaya sq., 7,
Moscow 105037, Russian Federation

Abstract. This paper describes a system for automatic emotion recognition developed to enhance the communication capabilities of an anthropomorphic robot. Two versions of the classification algorithm are proposed and compared. The first version is based on a classic approach requiring the action unit estimation as a preliminary step to emotion recognition. The second version takes advantage of convolutional neural networks as a classifier. The designed system is capable of working in real time. The algorithms were implemented on C++ and tested on an extensive face expression database as well as in real conditions.

Keywords: Service robotics · Emotional state · Face recognition · Deep learning · Convolutional networks · Action units

1 Introduction

Human emotion recognition systems have recently become one of the most prominent technologies in the field of human-robot interfaces. Such systems are considered useful for maintaining an effective communication process between human and anthropomorphic robot (human-like robot with skin imitation) since emotional reaction is one the most natural communication channels used by people. Providing that an anthropomorphic robot can recognize the emotional state of its human interlocutor, much more "user-friendly" scenarios can be implemented to make the conversation livelier and less tense for the human. The following interaction options can become possible:

- emotion copying: provided with face elements actuators, a robot could model the emotions it observes on faces of surrounding people [14];
- adaptive learning: a robot can collect a database of human emotional responses to its different actions and adjust its behavioral scenarios according to what is considered more favorable by the public [16];
- advanced service robotics: the robot can express its attitude to incoming messages from operator, as well as to its own state or environmental changes [17].

© Springer International Publishing AG 2017
A. Ronzhin et al. (Eds.): ICR 2017, LNAI 10459, pp. 20–28, 2017.
DOI: 10.1007/978-3-319-66471-2_3

Rapid development of face recognition techniques as well as significant increase of computational power of portable computing devices have made it possible for recognition systems to operate in real time and perform reasonably well to distinguish at least the basic emotions of the human face. There are several information sources that can be utilized and combined with each other to perform emotion classification, such as face image, body movements, voice recognition [18] and even brain neural activity [19]. However, this work is dedicated to face image recognition approach only. Next section describes the major research and developments in this field.

2 Related Work

Lately a number of commercial products and publications on emotion recognition have been issued [1–3]. Most researchers make use of the works published in 1970's by Ekman as a basis for quantitative description of facial expression. In his famous work called Facial Action Coding System (FACS) Ekman introduced the classification principle, which assumes 6 basic emotions: happiness, anger, surprise, fear, disgust and sadness (in some versions contempt is also addressed as a basic emotion, but this is an arguable statement) [4]. In FACS an emotion is referred as a combination of action units (AUs), which, in their turn, describe the specific face muscle contractions and the resulting effect on face appearance. Most researchers as well focus on obtaining the action unit scores as a preliminary step to emotion rate calculation. For example, in [7] extensive decision trees were constructed to pass from action unit rates to emotion rates, depending on specific ethnic characteristics of a subject's face. An approach which seems much similar to the one discussed in our work is presented in [14]: a set of action unit values is calculated using local image processing with Gabor filters and Dynamic Bayesian Network classification. An emotion is described as a combination of AUs. However, a very limited number of AUs is used in this work and using network at the early processing stages may cause the solution to be unstable under varying imaging conditions. Local geometric feature extraction (without direct connection to AUs) is performed in [15] to form a feature space for Kohonen self-organizing maps classifier training.

Another popular approach takes advantage on holistic image processing techniques combined with machine learning classifiers to extract the emotion rates from input images directly avoiding the AU calculation step. For example, in [12] Gabor filters are used as feature extractors followed by AdaBoost method to reduce the excessive feature subspace. An MLP neural network is trained to classify the feature vectors by corresponding emotion values. The authors claim to have reached over 98% recognition rate on JAFFE and Yale databases, however, such rate could be a result of MLP overfitting as the listed databases are relatively small and unrepresentative in the means of subject variability. In [13] histogram of oriented optical flow is used to form a bag of words on which a discrete hidden Markov Model is trained to predict the emotions. In [8, 9] spatiotemporal local binary patterns (LBP-TOP) were used as basic features. However, a whole sequence of frames is required for these approaches.

A serious obstacle for emotion recognition system developers including the mentioned above is the lack of open-source labeled databases, including so-called "in the wild" ones (with unconstrained environment conditions and camera characteristics). One of the most prominent available database designed specifically for emotion recognition applications is Cohn-Kanade database [5, 6], but it contains only grayscale images of generally rather mediocre quality. This limits the utilization of state-of-art deep learning approaches, while common feature-based methods often suffer from high vulnerability to head pose, lighting conditions and image resolution and quality. One more specific problem is that despite general classification rules had been figured out by experts, emotion display is a highly individual feature, which varies greatly between the subjects.

In this work we propose and compare two different approaches to the problem of emotion recognition: a feature-based classifier and a convolutional neural network. We tested our system on Cohn-Kanade database as well as on our own extensive emotion image database to achieve as much reliable results as possible.

3   Classification

3.1   Feature-Based Classifier

In this algorithm we take advantage of Ekman's AU description and focus on detecting their presence on face and calculating the score to characterize their intensity. As we assumed working with single frontal images, we have chosen 20 action units, which are visible on the frontal view. These include some geometric features, such as brow raise, mouth curve orientation as well as textural ones, such as wrinkles along forehead, nose sides and mouth corners. This approach requires a reference image of a subject in "neutral" state so that we can take into account the general appearance characteristics of a subject.

On the first step, we labeled our images with 68 landmarks using DLib library [10]. The geometric features can be calculated straightforward using the landmark positions, but this can lead to significant errors as landmark detectors tend to work quite unstable under the conditions of varying face expression and head pose. That is why we decided to use landmarks only to specify the regions of interest (ROIs) for each action unit. Different image processing techniques were then applied to each ROI. For example, we convolved forehead area image with horizontally oriented Gabor filter to calculate the large forehead wrinkles intensity. Smaller wrinkles were detected by applying standard deviation filter to correspondent ROIs. Specific nonlinear rank filtration and blob processing were used to estimate the mouth curve and corners orientation (Fig. 1).

After all AU scores are calculated, the next step is to normalize their scores to a certain common scale range. However, most of the scores may vary greatly from subject to subject which makes it impossible to build a reliable normalization model for the whole set of subjects. To deal with this obstacle we used a combined approach to normalization utilizing both statistical and subject-special parameters. As mentioned above, for each subject we obtained a reference image with neutral expression and calculated its AU scores. These values were used to calculate the relative scores for all other images for the given subject. These scores are once again normalized by mean and standard deviation values calculated across all of the images for all subjects in the

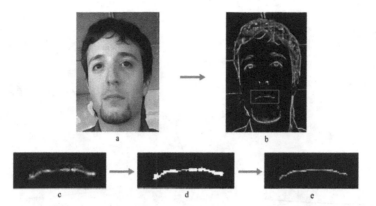

Fig. 1. Mouth curve estimation: a – initial image, b – rank filtering output, c – mouth ROI, d – binarization and small fragments filtering, e – contour approximation by curve

database. The normalization function was chosen to maximize the statistically significant deviations from mean (neutral) values and minimize the influence of outliers (artifacts) and values statistically close to means (Fig. 2):

$$
x_{normed} = \begin{cases} \tan^{-1}\left(\frac{x-\mu}{c-\mu}\right) \cdot \frac{2}{\pi} - 1, & \text{if } x \leq \mu, \\[2mm] -\left(\frac{x-c}{c-\mu}\right)^2, & \text{if } \mu < x < c, \\[2mm] -\left(\frac{x-c}{high-c}\right)^2, & \text{if } c < x < high, \\[2mm] \tan^{-1}\left(\frac{x-high}{high-center}\right) \cdot \frac{2}{\pi} - 1, & \text{if } x \geq high, \end{cases} \tag{1}
$$

where: $\mu = \frac{1}{N}\sum_{i=1}^{n} x_i$ – mean AU value among all samples from database; $c = \mu + \sigma$; $\sigma = \mu + \sqrt{\frac{1}{N-1}\sum_{i=1}^{n}(x_i - \mu)^2}$ – standard deviation of AU values; $high = \mu + 2\sigma$.

It is important to point out that some AUs are bidirectional: for example, brows can be lifted up or down from neutral position. This gives us two separate features in terms of classification, each with its own statistical moments. For calculation convenience, such AUs were split into two separate ones each with its own score (as shown on Fig. 2).

Using normalized AU scores one can decide whether one of six basic emotions (or neutral state) is present on a face by directly applying Ekman's decision rules [4]. However, direct implementation of these rules may lead to quite low sensitivity of the system since the error in one single AU would likely ruin the whole decision making function. In this work we used a probabilistic classifier with an option of solving "conflicting" outputs.

In this classifier each of the basic emotion Ei (i = 1…6) is assigned with its probability value:

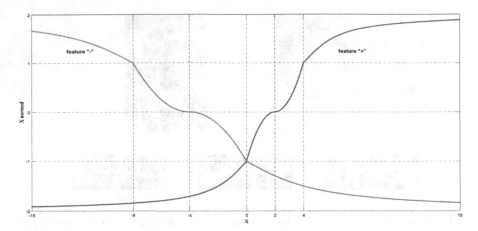

Fig. 2. Normalization functions for bidirectional AU: $x_{normed} = f(x)$ with $\mu = 0$, $\sigma_+ = 2$, $\sigma_- = 4$

$$E_i = \frac{\sum_{k=1}^{n} w_k x_k}{\sum_{k=1}^{n} |w_k|}, \tag{2}$$

where n is the number of AUs, x_k is the normalized score of k-th AU, w_k is the weight assigned to the k-th AU. Weights can have positive, zero or negative values, depending on whether an action unit is conducive, irrelevant or contrary to i-th emotion. Neutral expression corresponds to minimal weighted sum of all AU scores.

Some of the basic emotions are close to each other in terms of their composition, for example, some versions of the "fear" emotion consist of mostly the same features as the "surprise" one. However, even such different emotions as happiness and surprise have some common features, such as a slight squint and nose side wrinkles. In combination with some algorithm errors this may lead to a highly undesirable situation when several mutually exclusive emotions receive relatively high probabilities at the same time. We solved this problem by using an additional set of decision rules, which incorporates a special limited set of discriminative features for each "conflict" scenario to make a final choice of emotion. For example, mouth curve orientation is a discriminative feature for happiness vs disgust.

3.2 Neural Network Classifier

Convolutional neural networks have recently become a very popular approach in most of the applications related to image processing. The key to success of this approach is its versatility, a number of out-of-the-box implementations available and generally high performance comparing to other state-of-art methods [11]. One principal limitation of convolutional network is that it requires extensive and computationally expensive training on a high number of samples ("big data"). In this work we proposed a rather compact convolutional network consisting of three convolutional layers, and two fully-connected layers. Maximum pooling layer for dimension reduction follows each

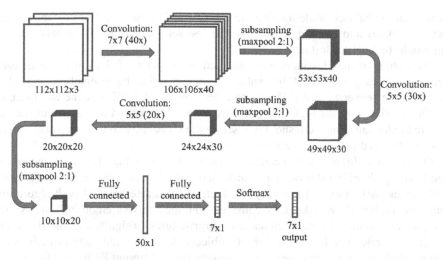

Fig. 3. The architecture of convolutional neural network

convolutional layer, and a dropout is used during training stage to improve the learning quality (Fig. 3). All the images were sorted into 7 classes (six basic emotions and neutral expression) and feeded to the algorithm regardless of which subject they belonged to. We also kept the images in RGB space to make use of color information.

Image normalization stage was required to align the samples to a uniform size and scale. Scaling, rotation and cropping of the images were performed based on eye center positions. Images were cropped and resized to the size of 112×112 pixels (54 pixels between eyes) and then normalized by mean and standard deviation values of pixel intensities independently for each color channel. To increase robustness of the classifier some images were randomly flipped from left to right, as well as some inconsistency was artificially added to the face ROI coordinates.

The algorithm was implemented in two different modes: with reference image and without. The first mode requires normalization using a reference image with neutral expression for the given subject; the second mode operates given only one current image.

4 Experimental Results and Discussion

Both algorithms described in previous sections were implemented on C++ and tested on an extensive database consisting of 3200 frontal images of 218 subjects. The images were taken under different lighting conditions and have different resolution (with minimal limitation of 60 pixels between eyes). The amount of images available for every emotion of a given subject varies from 0 to 10 (but there is always at least one neutral image for each subject). Overall amount of images per class is roughly the same to maintain the balance while training the classifiers. The subjects from database were randomly sorted into training, validation and test sets so that approximately 30% of data would be kept for testing.

The output of the classifiers was obtained as a vector of 7 elements with each element containing the probability value of corresponding basic emotion or neutral state, in range between 0 and 1. During test the "winner takes all" scheme was used, so that the emotion with the highest probability value was marked as the output one. In real-life implementation, a threshold was set on the probability so that several emotions could be declared present at the same time.

The results in form of confusion matrices are shown on Fig. 4. It is clear that the deep learning algorithm shows significantly better results than feature-based classifier. (73% versus 58% in overall accuracy). In fact, real characteristics of both algorithms would be higher if we take into consideration the "mixed emotions" states (for example, misclassification between fear and surprise brings a significant contribution to overall error rate, but for a number of subjects these emotions are actually quite similar). Both algorithms required the processing time of around 80 ms per frame on a computer with a 3.2 GHz Intel(R) Core(TM) i7 CPU and 16 GB RAM running using Windows 10, which is enough to run the system in real-time with frame rate of 10 fps.

Despite lower emotion recognition rate, a feature-based classifier has one certain advantage over a network approach: as previously described, it produces not only the emotion scores, but also the AU scores, which is quite useful output for some applications (such as an expression repetition scenario for human-robot communication). AU scores can also be calculated using independent convolutional networks for each of them, but this requires an extensively labeled image database and is more computationally expensive.

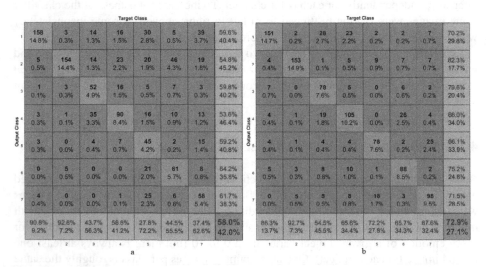

Fig. 4. Experimental results for a – feature-based algorithm and b – convolutional network algorithm. Classes: 1 – neutral; 2 – happy; 3 – sadness; 4 – anger; 5 – surprise; 6 – disgust; 7 – fear

5   Conclusion

The article describes a system for real-time automatic recognition based on the image processing algorithms. Two approaches were implemented and tested: a feature-based classifier and a convolutional neural network. The latter algorithm produced an average of 73% recognition rate which was considered effective enough to use it as a component for human-robot interface system. It was particularly used as a communication module for service robot "Alice" created by Neurobotics Ltd. In future, however, we plan to improve the quality and performance of a feature-based classifier as it provides a useful option of retaining the action units scores which can be utilized to create more versatile scenarios of human-robot communication.

Acknowledgements. This work was accomplished through financial support of Russian Foundation for Basic Research (RFBR), grant №16-07-01080.

References

1. Bartlett, M.S., Hager, J.C., Ekman, P., Sejnowskie, T.J.: Measuring facial expressions by computer image analysis. Psychophysiology **36**, 253–263 (1999). Cambridge University Press
2. Zaboleeva-Zotova, A.: The development of the automated determination of the emotions and the possible scope. Open Educ. **2**, 59–62 (2011)
3. den Uyl, M., van Kuilenberg, H.: The FaceReader: online facial expression recognition. In: Proceedings of Measuring Behavior 2005, 5th International Conference on Methods and Techniques in Behavioral Research, pp. 589–590. Noldus Information Technology, Wageningen (2005)
4. Ekman, P., Friesen, W.V., Hager, J.C.: Facial Acton Coding System: The Manual. Research Nexus Division of Network Information Research Corporation (2002)
5. Kanade, T., Cohn, J.F., Tian, Y.: Comprehensive database for facial expression analysis. In: Proceedings of the Fourth IEEE International Conference on Automatic Face and Gesture Recognition (FG 2000), Grenoble, France, pp. 46–53 (2000)
6. Lucey, P., Cohn, J.F., Kanade, T., Saragih, J., Ambadar, Z., Matthews, I.: The extended Cohn-Kanade dataset (CK+): a complete expression dataset for action unit and emotion-specified expression. In: Proceedings of the Third International Workshop on CVPR for Human Communicative Behavior Analysis (CVPR4HB 2010), San Francisco, USA, pp. 94–101 (2010)
7. Chandrani, S., Washef, A., Soma, M., Debasis, M.: Facial expressions: a cross-cultural study. In: Emotion Recognition: A Pattern Analysis Approach, pp. 69–86. Wiley (2015)
8. Huang, X., Wang, S., Liu, X., Zhao, G., Feng, X., Pietikäinen, M.: Spontaneous facial micro-expression recognition using discriminative spatiotemporal local binary pattern with an improved integral projection. J. Latex Class Files **14**(8), 1–13 (2015)
9. Zhao, G., Pietikäinen, M.: Dynamic texture recognition using local binary patterns with an application to facial expressions. IEEE Trans. Pattern Anal. Mach. Intell. **29**, 915–928 (2007)
10. King, D.E.: Dlib-ml: a machine learning toolkit. J. Mach. Learn. Res. **10**, 1755–1758 (2009)

11. Matsugu, M., Mori, K., Mitari, Y., Kaneda, Y.: Subject independent facial expression recognition with robust face detection using a convolutional neural network. Neural Netw. **16**, 555–559 (2003)
12. Owusu, E., Zhan, Y., Mao, Q.R.: A neural-AdaBoost based facial expression recognition system. Expert Syst. Appl. **41**, 3383–3390 (2014)
13. Kung, S.H., Zohdy, M.A., Bouchaffra, D.: 3D HMM-based facial expression recognition using histogram of oriented optical flow. Trans. Mach. Learn. Artif. Intell. **3**(6), 42–69 (2015)
14. Cid, F., Prado, J.A., Bustos, P., Nunez, P.: A real time and robust facial expression recognition and imitation approach for affective human-robot interaction using Gabor filtering. In: Proceedings of the IEEE/RSJ International Conference on Intelligent Robots and Systems (IROS), Tokyo, Japan, pp. 2188–2193 (2013)
15. Majumder, A., Behera, L., Subramanian, V.K.: Emotion recognition from geometric facial features using self-organizing map. Pattern Recogn. **47**(3), 1282–1293 (2014)
16. Zhang, L., Jiang, M., Farid, D., Hossain, M.A.: Intelligent facial emotion recognition and semantic-based topic detection for a humanoid robot. Expert Syst. Appl. **40**(13), 5160–5168 (2013)
17. Yuschenko, A., Vorotnikov, S., Konyshev, D., Zhonin, A.: Mimic recognition and reproduction in bilateral human-robot speech communication. In: Ronzhin, A., Rigoll, G., Meshcheryakov, R. (eds.) ICR 2016. LNCS, vol. 9812, pp. 133–142. Springer, Cham (2016). doi:10.1007/978-3-319-43955-6_17
18. Castellano, G., Kessous, L., Caridakis, G.: Emotion recognition through multiple modalities: face, body gesture, speech. In: Peter, C., Beale, R. (eds.) Affect and Emotion in Human-Computer Interaction. LNCS, vol. 4868, pp. 92–103. Springer, Heidelberg (2008). doi:10.1007/978-3-540-85099-1_8
19. Jirayucharoensak, S., Pan-Ngum, S., Israsena, P.: EEG-based emotion recognition using deep learning network with principal component based covariate shift adaptation. Sci. World J. **2014**, 10 p. (2014). Article ID 627892

Comparsion of Feature Extraction Methods for Finding Similar Images

Lukáš Bureš[(✉)], Filip Berka, and Luděk Müller

Faculty of Applied Sciences, New Technologies for the Information Society,
University of West Bohemia, Univerzitní 8, 306 14 Plzeň, Czech Republic
{lbures,berkaf,muller}@ntis.zcu.cz

Abstract. In this paper, we compare four methods of feature extraction for finding similar images. We created our own dataset which consists of 34500 colour images of various types of footwear and other accessories. We describe the methods that we experimented with and present results achieved with all of them. We presented the results to 23 people who then rated the performance of these methods. The best–rated method is selected for further research.

Keywords: Image similarity · Feature extraction · Local binary patterns · Colour histogram · Histogram of oriented gradients · Convolution neural network

1 Introduction

Searching by image has become a very popular tool for finding information on the internet over the past few years. The advantage of searching by image is that one does not need to know any keyword. A digital photo of the thing of interest is all one needs.

The usage of this tool could be looking for a name of certain goods. The tool should then provide the name and possibly a website where the goods can be purchased. Other example could be looking up information about a sight then providing the user with the name of the city where the sight is located. Other exemplary uses could be looking for: paintings, food and recipes, actors, animal/flower species, etc.

The problem of finding similar images in large datasets has been addressed many times e.g. [2,8]. In [8] they proposed a new method of similar image retrieval by using a joint histogram. A joint histogram is an alternative to colour histogram which incorporates additional information about the image by selecting other features. In their experiments the joint histograms outperformed the colour histograms on large databases.

Similar image search can be also helpful in medical image diagnosis and treatment planning. In [6] they investigated a method of retrieving relevant images as a diagnostic reference for breast masses on mammograms and ultrasound images.

© Springer International Publishing AG 2017
A. Ronzhin et al. (Eds.): ICR 2017, LNAI 10459, pp. 29–36, 2017.
DOI: 10.1007/978-3-319-66471-2_4

2   Methods

In this section, we will mention methods that we used for our experiments in Sect. 3.

2.1   Convolutional Neural Network

In 2012 Krizhevsky et al. presented a paper, titled "ImageNet Classification with Deep Convolutional Networks" [3]. This paper is regarded as one of the most influential publications in computer vision. They created a convolutional neural network that won the 2012 ImageNet Large-Scale Visual Recognition Challenge (ILSVRC) with top 5 error rate of 15.4%. The network was later called AlexNet. The architecture of AlexNet is depicted in Fig. 1, it has 5 convolutional layers and 3 fully connected layers. The last layer served the purpose of classifying input images into 1000 various classes.

The most important advantage of using AlexNet is no need to choose what features to extract from the training images. The convolution kernels change during the training to be able to extract features discriminative enough for the given problem. The feature vectors can be obtained from the second to last layer of the network.

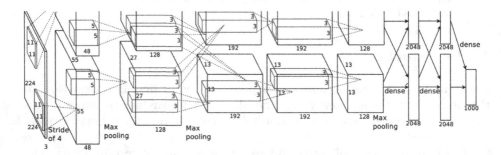

Fig. 1. Architecture of AlexNet implemented on 2 GPUs

2.2   Colour Histograms

For humans colour provides a not neglectable information about what they see. One way of using the colour information as features for machine learning algorithms are histograms. Histogram is an estimate of the probability distribution of a continuous variable. It was introduced in 1895 by Pearson [9]. A colour histogram is a representation of the distribution of colours in the image. For example a colour histogram of a grayscale image would be a set of 256 numbers (for 8-bit image) each corresponding to a quantity of that number in the image.

There are several colour spaces that one can take into consideration. For example there is grayscale which has only one so called channel in this case scales of gray. The other well known colour spaces are RGB, HSV, HSL, CMYK, YPbPr. These have more than one channel. If one wants to use colour histograms as a feature extraction method one way to obtain the vectors is computing the histogram for each channel separately and then simply concatenating them together.

2.3 Histogram of Oriented Gradients

Histogram of Oriented Gradient (HoG) was firstly introduced in 1986 by McConnel [4]. However it became widespread in 2005 by Dalal and Triggs when they used it with SVM classifier for pedestrian detection [1]. Since then HoG descriptors have been used for various object detection applications.

At first the method computes gradient using 1 dimensional centred derivative mask in both directions. The input image is then divided into cells for which histograms of gradient are computed. These histograms are usually normalized to achieve invariance of the descriptor to brightness transformations (using typically L^1 norm, L^2 norm or L^2–Hys norm). The cells are then grouped together into larger blocks. These blocks usually overlap. For each block the normalized cell histograms are averaged. Then the block shifts and the process is repeated. The final descriptor is obtained by concatenating the averaged block histograms.

2.4 Local Binary Patterns

The Local Binary Patterns (LBP) were introduced in 1996 Ojala in paper [7]. It was originally proposed for texture classification. It is a simple but efficient descriptor and it is invariant to brightness changes and rotation.

At first the input image is divided into cells. For each pixel of each cell the LBP code is computed. The LBP code is computed by comparing a pixel with its neighbours. If the center pixel has greater value than the neighbour a 1 is written to the LBP code. Otherwise a 0 is written. In the simplest form of LBP this returns 8–bit binary number which is usually converted to decimal. Then normalized histogram over each cell is computed. The final descriptor is obtained by concatenating these histograms.

2.5 Fast Library for Approximate Nearest Neighbours

The Fast Library for Approximate Nearest Neighbours (FLANN) was presented in 2009 by Muja and Lowe in paper [5]. It is a tool for solving nearest neighbour matching in high–dimensional spaces popular mainly for its speed. There is often no better algorithm for finding the nearest neighbour for those high–dimensional spaces than simple linear search. For many applications linear search is way too costly. That is the reason behind the need for algorithms that perform approximate nearest neighbour search. These return non–optimal neighbours but can be much faster than the exact search while still being accurate enough.

It can provide automatic selection of index tree and parameter based on the user's optimization preference on a particular data–set. They are chosen according to the user's preferences on the importance between build time, search time and memory footprint of the index trees. The type of index trees are KD, randomized KD and Hierarchical K–Means. It memorizes all the data from the input dataset. Then it automatically chooses the best nearest neighbour algorithm and parameters for the dataset.

3   Experiments

In this section, we briefly describe the dataset that we collected and how we used the methods mentioned in Sect. 2. The pipeline of how the first part of the system works, in general, is shown in the Fig. 2. All images are input to the feature extraction method which returns feature vectors. The FLANN mentioned in Subsect. 2.5 then memorizes all of them and chooses the best nearest neighbour algorithm.

All images are processed and saved by FLANN and then we can input a single image to the system and extract its feature vector. Then FLANN can find an image whose feature vector is the nearest neighbour of the input image feature vector. The pipeline of the second part of the system can be seen in Fig. 3.

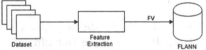

Fig. 2. The pipeline shows how the first part of the system works. The images are processed by a feature extraction method. Feature vectors are then sent to FLANN

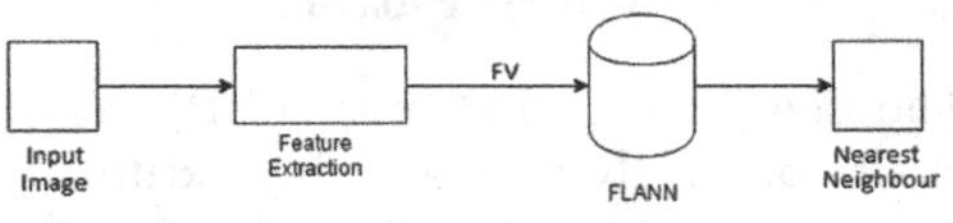

Fig. 3. The pipeline of the similarity recognition systems' part. The images are processed by a feature extraction method. Feature vectors are then sent to FLANN which finds selected number of the nearest neighbours

3.1   Dataset

We test our methods of finding similar images on a dataset that we created. The dataset contains 34500 colour images and we selected only images with extensions jpg, jpeg, and png. This leads to the final number of 33167 images.

The dataset[1] contains many different kinds of mainly footwear (eg. high heels, pumps, loafers, boots, sandals, hiking boots, sneakers, slippers, rubber boots, flip flops, and etc.) and its minor part are other accessories like wallets, handbags, bags, and etc. All these images we have found and downloaded from Google.

---

[1] https://drive.google.com/open?id=0Bzz60niNx1e1aDR4Sld1dzF3RWM.

We normalised all of the images to the size 227×227 because they did not have the same size. The 3D orientation of the footwear in the images is highly variable. The examples of images from the dataset are shown in Fig. 4.

Fig. 4. Examples of images in the dataset

3.2 Using AlexNet

In the first experiment, we use AlexNet (see Sect. 2.1). Its purpose here contrary to the original idea was not to classify the input image into 1000 classes but just to extract feature vectors. We decided to cut the pre-trained network so that the fully connected layer (fc6) would be the last one as can be seen in Fig. 5. This layer returns 4096–dimensional feature vector representing the image.

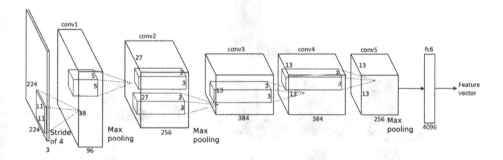

Fig. 5. AlexNet architecture with the fully connected layer fc6 as its last layer

3.3 Using Colour Histograms

In the second experiment, we use colour histograms (see Sect. 2.2) as feature vectors. We take each image and split it into its 3 RGB channels. For each channel we compute histogram. We obtain the final $256 \cdot 3 = 768$–dimensional feature vector by concatenating the histograms in BGR order.

3.4 Using Histograms of Oriented Gradients

In the third experiment, we use HoG (see Sect. 2.3) as feature vectors. We take each image and convert it to grayscale. Then HoG is computed with 16 orientations (bins) over one cell whose size is equal to the size of the image. The final 16–dimensional feature vector is returned.

3.5 Using Local Binary Patterns

In the final experiment, we use LBP (see Sect. 2.4) as feature vectors. We take each image and convert it to grayscale. Then LBP is computed for each pixel of the image. From LBP codes a histogram is computed and normalized. The final 26–dimensional feature vector is returned.

3.6 Evaluation

As we use methods each returning feature vectors with different dimensionality there is no easy way of automatically comparing success rates of these methods in the task of finding the most similar images. Because of that, we let people do the evaluation. We created a test where we selected 50 images of footwear as the input into the system. To the test, we chose 5 most similar images. We focused on finding footwear inner–class similarity that is why we did not select the other accessories mentioned in Sect. 3.1. We shuffled the order of the methods for each page so that people would not recognize any pattern. One page of the test can be seen in Fig. 6. The task was to assign a rating from 1 to 4 to each method, 1 being the best and 4 being the worst.

Fig. 6. One page of the test. The input image is at the top of the page. Each row represents a method of feature extraction. There is a handbag in the second row. That could serve as a hint that this method performed badly

4   Experimental Results

We assume that every person has a bit different opinion on how to compare similarity in images. Our assumption was that method based on neural network outperform other methods in our proposed test.

We gave the test to 23 people in total of which 15 were men and 8 were women. In the Table 1 we show mean rating of each method (the lower rating the better method), variance, and standard deviation

Table 1. Table shows average rating of each method. The lower rating the better method

Method	AlexNet	Colour Hist	HoG	LBP
Mean	**1.04**	2.92	2.95	3.09
Variance	**0.08**	0.73	0.64	0.70
Std	**0.28**	0.86	0.80	0.83

One can easily see that according to the people AlexNet features beat the hand–crafted ones almost flawlessly. The other methods performed more or less the same.

5   Conclusion

We compared feature extraction methods for finding similar images. We used AlexNet machine learned features, Colour histograms, Histograms of Oriented Gradient and Local Binary Patterns. Our experiments showed that the AlexNet performed the best and other methods performed about the same. The main reason for this is that there was no need to tune parameters with AlexNet. That was unfortunately not the case for the hand–crafted features.

We proposed dataset which contains 34500 colour images. It consists mainly of footwear on white background. This dataset can be easily used e.g. for segmentation task or for finding the n–most similar images as we used it.

In future work, we will focus mainly on feature extraction methods based on neural networks. Because nowadays it is very hard to overcome neural networks methods by hand–crafted features. Next, we will focus on extending, improving and annotating our dataset.

Acknowledgment. This publication was supported by the project No. LO1506 of the Czech Ministry of Education, Youth and Sports, and by grant of the University of West Bohemia, project No. SGS-2016-039.

Computational resources were supplied by the Ministry of Education, Youth and Sports of the Czech Republic under the Projects CESNET (Project No. LM2015042) and CERIT-Scientific Cloud (Project No. LM2015085) provided within the program Projects of Large Research, Development and Innovations Infrastructures.

References

1. Dalal, N., Triggs, B.: Histograms of oriented gradients for human detection. In: 2005 IEEE Computer Society Conference on Computer Vision and Pattern Recognition (CVPR 2005), vol. 1, pp. 886–893, June 2005
2. Jégou, H., Douze, M., Schmid, C.: Improving bag-of-features for large scale image search. Int. J. Comput. Vis. **87**(3), 316–336 (2010)
3. Krizhevsky, A., Sutskever, I., Hinton, G.E.: Imagenet classification with deep convolutional neural networks. In: Pereira, F., Burges, C.J.C., Bottou, L., Weinberger, K.Q. (eds.) Advances in Neural Information Processing Systems 25, pp. 1097–1105. Curran Associates, Inc., Red Hook (2012)
4. McConnell, R.: Method of and apparatus for pattern recognition, 28 January 1986. US Patent 4,567,610
5. Muja, M., Lowe, D.G.: Fast approximate nearest neighbors with automatic algorithm configuration. In: International Conference on Computer Vision Theory and Application VISSAPP 2009, pp. 331–340. INSTICC Press (2009)
6. Muramatsu, C., Takahashi, T., Morita, T., Endo, T., Fujita, H.: Similar image retrieval of breast masses on ultrasonography using subjective data and multidimensional scaling. In: Tingberg, A., Lång, K., Timberg, P. (eds.) IWDM 2016. LNCS, vol. 9699, pp. 43–50. Springer, Cham (2016). doi:10.1007/978-3-319-41546-8_6
7. Ojala, T., Pietikainen, M., Harwood, D.: A comparative study of texture measures with classification based on feature distributions. Pattern Recogn. **29**(1), 51–59 (1996)
8. Pass, G., Zabih, R.: Comparing images using joint histograms. Multimedia Syst. **7**(3), 234–240 (1999)
9. Pearson, K.: Contributions to the mathematical theory of evolution. Philos. Trans. R. Soc. Lond.: Math. Phys. Eng. Sci. **186**, 343–414 (1895)

Empowerment as a Generic Utility Function for Agents in a Simple Team Sport Simulation

Marcus Clements[✉] and Daniel Polani

Adaptive Systems Research Group, School of Computer Science,
University of Hertfordshire, Hatfield, UK
m.clements@mxcog.com, d.polani@herts.ac.uk

Abstract. Players in team sports cooperate in a coordinated manner to achieve common goals. Automated players in academic and commercial team sports simulations have traditionally been driven by complex externally motivated value functions with heuristics based on knowledge of game tactics and strategy. Empowerment is an information-theoretic measure of an agent's potential to influence its environment, which has been shown to provide a useful intrinsic value function, without the need for external goals and motivation, for agents in single agent models. In this paper we expand on the concept of empowerment to propose the concept of *team empowerment* as an intrinsic, generic utility function for cooperating agents. We show that agents motivated by team empowerment exhibit recognizable team behaviors in a simple team sports simulation based on Ultimate Frisbee.

Keywords: Empowerment · Intrinsic motivation · Artificial intelligence · Information theory

1 Introduction

Player behavior in team sports is a complex interplay of tactics and strategies that remains unpredictable and opaque despite the finely grained analysis of scientists, television pundits and millions of sports fans every weekend worldwide. Commercial computer sports simulations and RoboCup soccer teams adopt a "top-down" approach to analysis of the game, breaking down the strategic and tactical behaviors observed in team sports to construct highly complex ontologies of action and reaction [10]. Such ontologies can be used to devise tactics and strategies for human players, and with considerable effort, to devise heuristics for robot player actions. Humans can draw parallels with other sports, but a sophisticated model requires increased cognitive load [19], consuming a precious resource when decisions need to be made quickly. In the absence of explicit goals, or before an intricate system of goals and sub-goals has been developed through experience, humans and animals are nonetheless able take decisions and act, motivated by some internal value function. To mimic this phenomenon, Schmidhuber introduced "artificial curiosity" [20,21], a reward function based on the

© Springer International Publishing AG 2017
A. Ronzhin et al. (Eds.): ICR 2017, LNAI 10459, pp. 37–49, 2017.
DOI: 10.1007/978-3-319-66471-2_5

interestingness of the environment. Steels describes the "autotelic principle" [24], that proposes that agents may become self-motivated by balancing skills with challenges based on earlier work on "flow theory" by Csikszentmihalyi [7]. From dynamical system theory, *homeokinesis* [8] uses an adaptive model of a dynamic environment to predict actions. Subsequent work by Ay et al. [2] utilizes information theory to predict a future world state from a known history where, as with homeokinesis, the value of the predicted information is dependent on the depth of historical information. By adopting a "ground-up" perspective, in contrast to considering a complex set of game knowledge driven heuristics, we consider an *intrinsic motivation* measure based on *Empowerment* [16] for decision making in team game play that does not require prior tactical or strategic knowledge. Empowerment can provide a motivation for action in environments where little or nothing is known about the past, but it assumes that future options can be predicted from the current perceivable world state [18].

1.1 Empowerment

Empowerment is an information-theoretic measure of the *potential* of an agent to affect the world, using its actuators as detected by its sensors, and can be used as a generalized intrinsic utility function for individual agent behavior [14, 16]. Salge et al. [18] suggest that "Empowerment provides a task-independent motivation that generates AI behavior which is beneficial for a range of goal-oriented behavior". A colloquial expression for maximizing empowerment could be "keeping your options open". If we consider the agent's interaction with the world as a causal Bayesian network, we can discretize the agent's actions as a series of iterations of the perception-action loop, in which the agent perceives the world, and acts on the received percepts during each iteration. Empowerment is formalized as the channel capacity [23] of the information flow between the actions of an agent, and the effect of those actions some time later [18]. We calculate the empowerment from the world state a given number of iterations in the future, referred to as *n-step* empowerment [18]. Previous studies have demonstrated the utility of empowerment in various settings, including in a simple maze [16], a grid world with a pushable box [16], and as a method of stabilizing a dynamic system in the form of a pole balancing on a cart [15].

1.2 Team Empowerment

Kelly defines the *superorganism* as "a collection of agents which can act in concert to produce phenomena governed by the collective" [13]. Duarte et al. proposed the concept of a sports team as a superorganism [9]. The team is considered to be a single entity, with its own intrinsic motivation, formed by the collective implicit exchange of information between team members, through movement and spatial positioning. Much of the existing work on empowerment has focused on individuals, however for multiple agents in the same environment, the agents actions each influence the world, and thus their empowerment is dependent on the actions of others. Capdepuy et al. demonstrated that structure can emerge in scenarios with multiple agents, all independently motivated by empowerment [4–6]. In this paper we propose an approach to the intrinsic motivation of a collective of cooperating agents, introducing the concept of *team empowerment* as a generalized utility function for the team superorganism, in a simplified team sport simulation, based on Ultimate Frisbee. We show that player agents, in teams motivated solely by empowerment, exhibit recognizable team sport behavior.

2 Model

2.1 Ultimate Frisbee

Ultimate Frisbee [26] is a field game with players in two teams of four to eight players throwing and catching a flying disc. The game is played on a rectangular field (see Fig. 1) with an area at each end, known as the end zones. For each point, play begins with the teams each in their own end zone. The team selects a player to start the point by throwing the disc downfield towards the opposition. As soon as a member of the opposing team takes possession of the disc, by catching or picking it up once landed, normal mid-game play begins. The player with the disc may not move, but will attempt to throw the disc (pass) to a teammate. Opposition players will attempt to block or intercept passes. If the disc is caught by an opponent, or touches the ground the opposing team gains possession. The aim of the game is to catch the disc in the opposing end zone, which scores one goal for the team in possession. Once a goal has been scored, the teams swap end zones and play begins again. The first team to reach an agreed number of points wins the game.

2.2 Basic Tactics

Players of Ultimate Frisbee adopt basic tactics of movement and possession of the disc that are common to other team games, notably Association Football (Soccer). Players try to gain possession by intercepting an opponents pass. Once in possession of the disc they try to pass to a teammate. Players on the same team as the player in possession "cut" (run) to try to find space on the field away from other players so that they are more

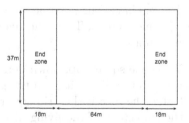

Fig. 1. Ultimate Frisbee field

likely to successfully receive a pass. Players on the opposing team try to intercept the pass, or make it more difficult for the player in possession by positioning themselves nearby and reducing the available throwing angles.

2.3   Intrinsic Motivation

Our model considers a drastically simplified model of Ultimate, where nonetheless essential aspects of game play can be observed, requiring only a minimum set of rules. The beginning of the game, the scoring of a point, and the restarting of the game after a point is scored, involve special states and rules. For simplicity, just the mid-game is considered, where play has begun, and the disc is on the ground, or already held by a player. Scoring goals, and the subsequent restart of the game, are not simulated in our current model. The reason for this is that we wish to illustrate the operation of empowerment as an intrinsic motivation per se, while scoring goals introduces an external "motivation" in the form of a traditional utility. While the combination of the latter with the utility-less empowerment formalism is the topic of current research, in the present paper we focus on how far the pure, task-less empowerment formalism can already shape team behavior. As we will see, empowerment as an intrinsic motivation leads to the emergence of recognizable team sports behaviors including passing and interceptions, but in this experimental model, it will not lead to goals. There are many ways in which goals could be added to the model, but this would blur the contribution of empowerment itself, which can achieve several desirable behaviors without the concept of explicit goals.

2.4   Representation

The model consists of a 20 × 20 grid of locations with two teams of two players per team, compared to five or more players per team in the full setup. The grid represents the central part of the playing field, and does not include the end-zones. Only the mid-game is modeled, goals will not be scored, so end-zones and goal detection is not required. The model runs in turns, with each agent having an opportunity to act in each turn. Each action takes one turn to complete.

The disc has four possible mutually exclusive states: Landed, Thrown, Flying, Held.

Players have four possible states: Static, Moving, Catching or picking up the disc (cannot move), Throwing (can move). The available actions for each agent each turn are:

1. Move one square up, down, left or right, or stop.
2. Pick up the disc if it is on the ground in the same square.
3. Catch the disc if the grid square in which the agent is located is crossed by the trajectory of the flying disc.
4. Throw the disc to any square on the field within the configurable maximum throwing distance.

The model has the following constraints for the players based on the rules of the game:

1. Cannot move outside the field.
2. Cannot move onto the same square as another player.

3. Cannot move when holding the disc (as per Ultimate Frisbee rules).
4. Cannot take possession of the disc again after throwing until another player has thrown (as per Ultimate Frisbee rules).

Once the disc has been Thrown, in the next turn it will be Flying. The disc has a velocity configurable as the number of squares it will travel per turn once thrown. Once the disc has been flying for sufficient turns to reach its destination, it will be Landed unless it is caught by a Player while Flying in which case it is Held. As this is a two-dimensional model, the trajectory of the disc has no height, so it may be caught by a player at any point on its path. Figure 2 shows the state diagram for players throwing and catching the disc. Each player (with number n) has a location vector (x_n, y_n) denoting a cell on the grid. The disc itself has a location vector (x_d, y_d), and a destination vector (x_{dest}, y_{dest}) which is only relevant when it is in the Thrown or Flying state. The state of the model (S_m) in any given turn can be represented by the locations and state (s_p) of the players, and the location, state (s_d), and destination of the disc.

$$S_m := D[x_d, y_d, s_d, x_{dest}, y_{dest}], P_0[x_0, y_0, s_0], \ldots, P_n[x_n, y_n, s_n] \qquad (1)$$

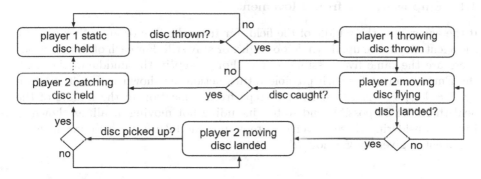

Fig. 2. State transitions for players and disc during throwing and catching action sequences

2.5 Player Precedence for Catching and Collisions

Players may not occupy the same square in the grid. During each turn the list of players is reordered according to a pre-determined, pseudo-random sequence where player positions in the list are evenly distributed over many iterations. Potential collisions are avoided because players take action in list order. If more than one player is located on the trajectory of the disc in a given turn, the closest to the disc origin may catch the disc.

3   Calculating Empowerment

In the general case, empowerment is computed via the channel capacity of the action-perception loop. This case captures partial information, stochastic and both continuous and discrete-state models. In our case, however, we will assume a discrete, deterministic, full-information model or convert it into such, where necessary, drastically simplifying the computation of empowerment, while retaining the expressivity of the phenomena. This simplifies the empowerment calculation to computing the logarithm of the number of distinct states reachable in n steps, which will be measured in bits, since we will use the binary logarithm. For the derivation of the n-step empowerment calculation from channel capacity please refer to [16]. As empowerment is essentially reduced to counting the end states in our deterministic scenario, the logarithm is not strictly necessary here; however, we retain it in the formula, permitting expression of empowerment in bits and keeping the present results conceptually and quantitatively consistent with the formalism generalized to stochastic and partial information scenarios.

$$\mathfrak{E} = \log_2 |S_n| \, bits \tag{2}$$

3.1   Empowerment from Movement

If the agent is in the middle of the field, far from the disc or any other players, the agent may move up, down, left or right, or stay still. For each of those actions there are the same five possible actions. For 1 step in the middle of the field, the Empowerment for each possible future action, as shown in Fig. 3, is given by $\mathfrak{E} = \log_2 5 = 2.32 \, bits$. After 3 steps from a location in the middle of the field there are 25 possible end states, including not moving at all, as shown in Fig. 4. Considering three movement steps into the future, the empowerment of the agent is given by $\mathfrak{E} = \log_2 25 = 4.64 \, bits$.

3.2   Empowerment from Throwing the Disc

An agent holding the disc may throw it to any other location within the maximum throwing distance. Given that the field has 20×20 locations, and a large maximum throwing distance, there are 399 potential resulting states for the disc as a result of the throw, thus the empowerment of the agent holding the disc is given by $\mathfrak{E} = \log_2 399 = 8.64 \, bits$. Note that empowerment itself just assigns an intrinsic (pseudo-)utility value to the states under consideration. In the present model, this pseudo-utility induces behavior in the same way a traditional utility function would do; namely, in a given state, one chooses the action that changes the world state into a world state where the present agent has the largest value of empowerment (here: reachable states in the given lookahead horizon)[1]. The

---

[1] Note that this is not a utility in the conventional sense for reasons that require some subtle discussion which is beyond the scope of the present paper. For the present case, this distinction is immaterial.

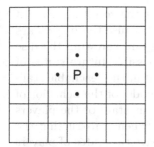 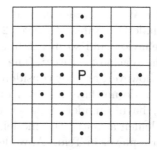 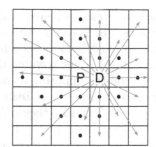

Fig. 3. Possible end states after one step. If the player at P does not move at all that is one possible end state. The dots show the reachable locations

Fig. 4. Possible end states after three steps. If the player at P does not move at all that is one possible end state

Fig. 5. Possible end states after three steps for player agent starting at P adjacent to disc D. The arrows show a subset of the possible disc trajectories

deterministic, full-information case, allows us to even select this action directly from the lookahead tree propagating from the current state (see below for an illustration).

Example 1. Agent in cell adjacent to disc.

As a more detailed example of determining the possible future states for an agent, let us consider a typical situation. The disc has landed in the adjacent square to an agent, and we are considering the empowerment of the possible next states given a three-step lookahead horizon. The agent has all the usual movement options at each step, which for two steps leads to 25 possible end states. If the agent steps onto the location of the disc, it has one additional option - to pick up the disc. Once the agent has the disc it may throw it to one of 399 possible locations. There are three groups of three-step action sequences available to the agent:

1. Move $\rightarrow$ Move $\rightarrow$ Move (25 end states)
2. Move $\rightarrow$ Move $\rightarrow$ Pick up disc (1 end state)
3. Move $\rightarrow$ Pick up $\rightarrow$ Throw disc (399 end states)

Moving onto the square with the disc in the first turn is the agent's future state with the highest empowerment as none of the other possible future states can lead to throwing the disc, therefore that is the action the agent will take. Figure 5 illustrates the possible end states for a player agent adjacent to the disc.

3.3 Team Empowerment

All agents on the same team are considered as a single super agent (from superorganism) for calculating team empowerment. For Player 1 and Player 2, on the same team, when we calculate the empowerment for the resulting state of an action of Player 1, we include the states of Player 2 because the players are able to affect each other's behavior, and so in catching and throwing the disc their states count towards the reachable states[2]. The possible future states of agents on the opposing team however, are not considered in the reachable state calculation. Players in Team A, do not have a forward model of the behavior of players in Team B, they only have a model of the current state of the players (their locations). In the implementation of the model, when constructing the search tree of future states for a player in Team A, we leave the states of players in Team B unchanged from step to step. In the absence of information about the behavior of the opposition, the agents assume that opposition players take no action.

3.4 Implementation of the Model

The Java implementation of the model employs an object-oriented architecture to represent the model state as an acyclic directed graph. Each vertex in the graph represents the state of the simulation, and each edge in the graph represents an action. At each time step the graph is constructed by a recursive function. After n steps (for n-step empowerment) the number of end states is counted for each action. The choice of action to take from the current state is made simply according to which action (edge) will lead to the most future options (terminal vertices). The findings were captured into testable scenarios, providing initial conditions and assertable outcomes, which are repeatable because the model is deterministic. A graphical representation of the model is presented in the user interface, showing the changing position of the players and the disc (Figs. 6 and 7). The implementation provides the most basic environment for the expression of empowerment-driven behavior. Considerable further work is required to develop the team empowerment formalism to a more realistic stochastic, continuous environment before optimizing the model for real-time performance.

4 Experiments and Results

We employed the empowerment principle in a number of scenarios, typical of Ultimate Frisbee, to investigate how far it, by itself, and without further externally specified utilities, would drive the agents to take tactically plausible actions, based on the rules of the game [26]. After observation of these actions[3], each result was confirmed by creating a repeatable test scenario inspecting the graph of states and actions generated during the empowerment calculation.

---

[2] It is possible to consider other friendly agents in a more general way, as still different from the egocentric agent, but nonetheless cooperative. This is research in progress.

[3] Video captures of the simulation can be viewed at: http://mxcog.com/research/team_empowerment.

4.1 Finding Space

When not holding the disc or catching, an agent will move away from the sides of the field. At the sides the empowerment is lower because the future range of states available by move actions is restricted. An agent will also move away from another agent for the same reason. The distance the agents maintain from the sides, or each other, increases as the lookahead increases. "Finding space" is a common basic tactic in Ultimate and other team games, particularly Soccer.

4.2 Retrieving the Disc

An agent that is within the lookahead number of steps of a disc on the ground, will move directly to the disc and pick it up. Being in possession of the disc allows the throwing actions and so has a drastically higher empowerment than leaving the disc on the ground. Retrieving the disc, once it has hit the ground after a wayward pass, is an essential behavior of in Ultimate [26], and equivalent to collecting the ball after an inaccurate pass in Soccer.

4.3 Passing

With a three-step lookahead (or higher), an agent in possession of the disc will throw the disc to a location where a teammate may catch it. Throwing the disc to a teammate is the action leading to a successor state with the highest empowerment, because the future states of the teammate are taken into account in the calculation. In the subsequent turn, the disc will be in the *Flying* state with a trajectory crossing the cell containing the teammate, and thus catchable. For the teammate, being in possession of the disc has a higher empowerment than the disc landing on the ground, so they will catch it. After catching the disc, an agent must wait until the next turn to throw. While throwing the disc, an agent may also move, but while catching they may not. During the experiments, agents on the same team repeatedly throw ("pass") the disc to each other, moving when not catching. Passing is an essential behavior in Ultimate [26] and other team sports including Soccer.

4.4 Interceptions

An agent located on the trajectory of the disc can catch the disc. Note that, to compute empowerment, agents need to have a forward model of what will ensue from taking particular actions, including the destination of the disc once thrown. During the empowerment calculation, the tree of possible future states will include catching the disc if the agent is on the trajectory of the disc. Since being in possession of the disc has a higher empowerment than not, the agent will catch the disc. Catching the disc, when a player on the opposing team has thrown it, is known as an "interception", and is an important behavior in Ultimate [26] and other team sports. Figures 6 and 7 show an attempted pass and subsequent interception taking place in the model.

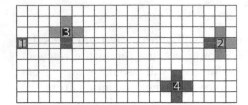

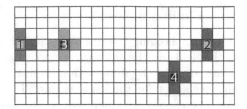

Fig. 6. Partial screen capture of the model running with a 3-step empowerment lookahead and a disc velocity of 20. Players 1 and 2 are on Team A, players 3 and 4 are on Team B. The grey line between the players shows the trajectory of the disc as thrown by Player 1 to Player 2. The darkness of the grey shading indicates the relative empowerment of that players possible subsequent states

Fig. 7. Partial screen capture of the model one time step after the screen capture in Fig. 6. Player 1 can now move and will have a slightly higher empowerment moving right because the edge of the eld constrains movement options. For Player 3 the highest empowerment state results from catching, so it has moved down and is now in possession of the disc

5   Discussion

We have shown that *team empowerment* as a generic utility function for agents in a sports simulation leads to emergent team game behavior in the form of four basic tactics: *finding space*, *retrieving the disc*, *passing* and *interceptions*. By the rules of Ultimate [26], if no player on the same team catches a thrown disc, possession passes to the opposition. A player intercepting a pass between opponents gains possession of the disc, increasing their empowerment. Thus catching the disc is a better strategy than not catching, and consequently stepping into the trajectory of the disc is a better strategy than not doing so. Moving away from opposition players makes a successful pass more likely by reducing the opportunity for an opponent to intercept. A human playing Ultimate Frisbee, with no prior knowledge of tactics or strategy, could employ a pseudo-empowerment strategy by acting to maximize future options. Our results indicate that a player utilizing such a strategy, would be a much more useful member of the team from the outset than a player merely choosing actions at random from the options allowed by the game. The parameters of the model automatically favor capturing and subsequently throwing the disc because throwing options are more numerous than moving options. It therefore does not seem surprising that these action sequences emerge from the empowerment dynamics. However, this is emphatically intended—empowerment is expressly a quantity that emerges from the interaction of agents with their environment, and this behavior emerges from it "naturally", rather than being externally imposed on the agents. We remind the reader that the present experiment is inspired by the box-pushing scenario from [16] where a manipulable box on a grid, merely by virtue of its affordance, creates an empowerment "attractor" around the box. This can be seen as a drastic abstraction of a "ball" attracting an agent by pure virtue of being able to be kicked around; in a way, an agent detecting that "playing football" with a box is "fun" [22]. For this to work, it is unavoidable for generic measures for intrinsic motivation to make heavy use of the environment. However, previous studies show that empowerment produces plausible and human-like behaviors in a variety of scenarios, including balancing tasks, puzzles, obstacle avoidance and more [18], with its completely generic approach, not utilizing a task-specific utility (and emphatically not trying to be somehow "best" according to some specialized criterion, but rather "reasonable" in many generic scenarios). This indicates that the environment itself is often sufficient to induce naturally expected behaviors in many cases from the outset, and that utilities typically imposed on such scenarios may just capture properties which are implicitly "encoded" in the scenario itself. Consequently, one tentative hypothesis is that humans would utilize principles which are similar in spirit to cope with the large regions of their task space without the guidance of dedicated and detailed utility functions [18]. Thus, our proof-of-principle empowerment-based studies in the present paper suggest a concrete and operational approach of how to address the generic interaction of an agent team in a ball-like game, in an intrinsic and natural way; this would happen in advance of specifying a detailed, problem-specific utility function, endowing first the agents with a set of "common sense" baseline behaviors, that could later be refined by a more specific reinforcement learning approach. Extensions of the empowerment formalism to the team level could take into account hierarchization techniques such as *Layered Learning* [25], by which "ritualized action sequences" (which could be considered an empowerment-induced analogy of the Reinforcement Learning concept of options [3]) could be compacted, and utilized as meta-actions, to serve for the empowerment computation at a higher level. Basic empowerment, as implemented here, will select the states with most options and not take account of any future development. To look beyond the empowerment horizon, one could adopt the "soft horizon" method from [1] which augments the empowerment computation by incorporating a "halo" of futures corresponding to the states reached by the selected action sequences. However, we have shown that already with basic empowerment the resulting strategies are capturing many of the natural "first approximations" to tactically meaningful behavior.

6   Future Work

We hypothesize that with a more sophisticated simulation, further recognizable Ultimate Frisbee tactics would emerge such as blocking passes and marking (restricting options) of opponents. There are many similarities in game play and tactics between Ultimate Frisbee and Soccer, and we anticipate similar results using team empowerment in a comparable model based on Soccer. Furthermore we hypothesize that, when combined with reinforcement learning, empowerment as an underlying strategy could lead to the emergence of longer term strategies such as formations and movement in anticipation of expected actions by other players on the team. For strategies such as this to emerge, it is necessary to master the complexity of more general computations of empowerment for the continuous and stochastic case, and for the case of imperfect team members, and actively antagonistic players. While a number of empowerment estimation approaches for continuous/stochastic scenarios in the single-agent case have been developed [11,12,17] the team aspect sketched above requires considerable extension, improvement and scaling up of the formalism. This will be the focus of future work, to harness the opportunities that the empowerment formalism offers to produce "intuitive" baseline team behavior.

Acknowledgements. DP would like to acknowledge support by the EC H2020-641321 socSMCs FET Proactive 713 project, the H2020-645141 WiMUST ICT-23-2014 Robotics project.

References

1. Anthony, T., Polani, D., Nehaniv, C.L.: General self-motivation and strategy identification: case studies based on Sokoban and Pac-Man. IEEE Trans. Comput. Intell. AI Games **6**(1), 1–17 (2014)
2. Ay, N., Bertschinger, N., Der, R., Güttler, F., Olbrich, E.: Predictive information and explorative behavior of autonomous robots. Eur. Phys. J. B-Condens. Matter Complex Syst. **63**(3), 329–339 (2008)
3. Barto, A.G., Mahadevan, S.: Recent advances in hierarchical reinforcement learning. Discret. Event Dyn. Syst. **13**(4), 341–379 (2003)
4. Capdepuy, P.: Informational principles of perception-action loops and collective behaviors. Ph.D. thesis, University of Hertfordshire (2010)
5. Capdepuy, P., Polani, D., Nehaniv, C.: Maximization of potential information flow as a universal utility for collective behavior. In: IEEE Symposium on Artificial Life, pp. 207–213. IEEE (2007)
6. Capdepuy, P., Polani, D., Nehaniv, C.L.: Perception-action loops of multiple agents: informational aspects and the impact of coordination. Theory Biosci. **131**(3), 149–159 (2012)
7. Csikszentmihalyi, M.: Beyond Boredom and Anxiety. Jossey-Bass, San Francisco (2000)
8. Der, R., Steinmetz, U., Pasemann, F.: Homeokinesis: a new principle to back up evolution with learning. Max-Planck-Institut für Mathematik in den Naturwiss (1999)
9. Duarte, R., Duarte, A., Correia, V., Davids, K.: Sports teams as superorganisms. Sports Med. **42**(8), 633–642 (2012)
10. Dylla, F., Ferrein, A., Lakemeyer, G., Murray, J., Obst, O., Röfer, T., Schiffer, S., Stolzenburg, F., Visser, U., Wagner, T.: Approaching a formal soccer theory from the behavior specification in robotic soccer. In: Dabnichki, P., Baca, A. (eds.) Computers in Sport, pp. 161–186. WIT Press, London (2008)
11. Jung, T., Polani, D., Stone, P.: Empowerment for continuous agent-environment systems. Adapt. Behav. **19**(1), 16–39 (2011)
12. Karl, M., Bayer, J., van der Smagt, P.: Efficient empowerment (2015, preprint)
13. Kelly, K.: Out of Control: The New Biology of Machines, Social Systems and the Economic World. Perseus, Cambridge (1994)

14. Klyubin, A., Polani, D., Nehaniv, C.: Empowerment: a universal agent-centric measure of control. In: The 2005 IEEE Congress on Evolutionary Computation, vol. 1, pp. 128–135 (2005)
15. Klyubin, A., Polani, D., Nehaniv, C.: Keep your options open: an information-based driving principle for sensorimotor systems. PloS ONE **3**, 12 (2008)
16. Klyubin, A.S., Polani, D., Nehaniv, C.L.: All else being equal be empowered. In: Capcarrère, M.S., Freitas, A.A., Bentley, P.J., Johnson, C.G., Timmis, J. (eds.) ECAL 2005. LNCS, vol. 3630, pp. 744–753. Springer, Heidelberg (2005). doi:10. 1007/11553090_75
17. Mohamed, S., Rezende, D.J.: Variational information maximisation for intrinsically motivated reinforcement learning. In: Neural Information Processing Systems (NIPS) 2015 (2015)
18. Salge, C., Glackin, C., Polani, D.: Empowerment–an introduction. In: Prokopenko, M. (ed.) Guided Self-Organization: Inception. ECC, vol. 9, pp. 67–114. Springer, Heidelberg (2014). doi:10.1007/978-3-642-53734-9_4
19. Samuelson, L.: Analogies, adaptation, and anomalies. J. Econ. Theory **97**(2), 320–366 (2001)
20. Schmidhuber, J.: Curious model-building control systems. In: IEEE International Joint Conference on Neural Networks, pp. 1458–1463. IEEE (1991)
21. Schmidhuber, J.: Exploring the predictable. Adv. Evol. Comput. **6**, 579–612 (2002)
22. Schmidhuber, J.: Formal theory of creativity, fun, and intrinsic motivation (1990–2010). IEEE Trans. Auton. Mental Dev. **2**(3), 230–247 (2010)
23. Shannon, C.E.: A mathematical theory of communication. Bell Sys. Tech. J. **27**, 623–656 (1948)
24. Steels, L.: The autotelic principle. In: Embodied Artificial Intelligence: International Seminar, Dagstuhl Castle, Germany, 7–11 July 2003, pp. 231–242 (2004). (Revised Papers)
25. Stone, P.: Layered Learning in Multiagent Systems: A Winning Approach to Robotic Soccer. MIT Press, Cambridge (1998)
26. World Flying Disc Federation Ultimate Rules Committee: WFDF Rules of Ultimate 2017 (2017). http://www.wfdf.org/files/WFDF_Rules_of_Ultimate_2017_-_FINAL_-_31_Dec.pdf

Survey of Modular Robots and Developed Embedded Devices for Constructive and Computing Components

Vladimir Dashevskiy, Viktor Budkov, and Andrey Ronzhin^(✉)

SPIIRAS, 39, 14th line, St. Petersburg 199178, Russia
ronzhin@iias.spb.su

Abstract. The specificity of modular robotics is the ability to connect individual robots to reconfigurable complexes with space structures. Approaches suitable for reconfiguring a swarm of modular robots were analyzed during the preparation of the survey. The analysis of projects on modular robotics in the publications of the last ten years has also revealed the prospects of attracting cloud robotics techniques for distributed sensory data collection, management and centralized training using stationary computing resources. The developed embedded devices for sensor data processing and control of servos of modular robots are described.

Keywords: Modular robotics · Embedded devices · Servos · SMARC · Swarm robotics · Cloud robotics

1 Introduction

The fundamental problem that arises in developing algorithms for group interaction of robotic complexes in modular, swarm and cloud robotics, with the contact connection of homogeneous robotic elements to unified structures in a three-dimensional space, is associated with the limited capabilities of individual robots, due to small overall dimensions, low energy reserves, etc. When reconfiguring a robot swarm, algorithms for controlling robotic pairing of robots are necessary both during the formation of the three-dimensional structure and after completion to maintain the shape of the swarm of the required configuration.

The most important specificity of modular robotics (MR) is the solution of kinematics problems of rigidly interconnected elements in conditions of limited built-in resources necessary for communication, connection and movement of modules, when performing reconfiguration tasks, assembling, capturing, moving a set of modular robots. The concept of robots with a morphologically variable structure was proposed by Toshio Fukuda in 1988 and was later called modular robotics [1], the goal of which is to overcome the limitations on flexibility and adaptability inherent in traditional robots with a fixed spatial structure. Modular robots consist of separate blocks (modules) having several degrees of freedom and mechanisms of interconnection, which provide the configuration of the robot's structure for the execution of the current applied task. In the work [2], several basic tasks, solved by MR, are distinguished: reorganization in a

© Springer International Publishing AG 2017
A. Ronzhin et al. (Eds.): ICR 2017, LNAI 10459, pp. 50–58, 2017.
DOI: 10.1007/978-3-319-66471-2_6

surface, a given form; movement; manipulation with external objects; support and balancing of unstable objects. To solve the above tasks, the MR has the following 9 main functions: (1) self-reconfiguration; (2) movement in the form of overflow; (3) movement in the form of a gait; (4) self-assembly; (5) self-disassembly; (6) self-adaptation; (7) capture of the object; (8) group action; (9) the environment of the object.

The formation of modular structures is made on the basis of groups and a swarm of robots. Algorithms of swarm behavior in most cases were taken from analogies with biological prototypes (a swarm of bees, a flock of birds, fish, insects, etc.) and are associated with finding food, moving in a flock, moving to light, synchronizing movement. To model the algorithms of swarm behavior, a much larger number of units of used robots is required, so the cost factor is one of the main factors in carrying out full-scale experiments. In the work [3], the analysis of the cost of individual robots is given; the original architecture of a cheap miniature robot is proposed; the results of the movement experiments of a group of one hundred robots, equipped with the simplest functions of movement based on a piezo motor and determination of the distance using an infrared transceiver, are shown.

From the point of view of the paradigm of cyberphysical systems, mobile robotic complexes are cyberphysical devices interacting remotely over the network together with other fixed and embedded means. The development of cloud technologies had a serious impact on robotics, resulting in new scientific directions, such as network robots [4], cloud robotics [5], cyberphysical robotics [6]. In cloud robotics, service-oriented architectures (SOA) have been used extensively in cyberphysical systems, but in this case physical devices, functioning and interacting with humans, are mobile robots that have a greater degree of autonomy and mobility than mobile client devices or embedded computerized means of an intellectual space. A new type of the architecture was proposed: robot as a service (RaaS) – by analogy with SOA, adopted in cloud technologies.

In the work [7], the main trends in the development of modern robotics are described. There are 10 classes of robots which are actively being developed: (1) bioinspiration robots using solutions similar to biological prototypes; (2) mobile micro-, nano-, femtorobots for solving medical problems; (3) walking machines, including humanoid robots; (4) robotic toys for educational and entertainment purposes; (5) ubiquitous robots, based on the paradigm of intelligent space, where each robot is accessible over the network, has a friendly user interface and is able to provide the necessary service at any time on the basis of the analysis of the current context; (6) home service robots for cleaning the house and surroundings; (7) cloud robots that use distributed external computing resources to process sensory information; (8) unmanned aerial vehicles having a different degree of autonomous control (from remote-controlled to fully autonomous), different number of engines and requirements for the runway; (9) autonomous transport systems with five levels of automatism ("No automation" – controlled by a person; "Driver Assistance" – with automatic means of helping the driver; "Partial (Monitored) Automation") – with partial automation, where driving is partly performed by automatic driving system (ADS); "Conditional Automation" – when ADS has full control in certain circumstances; "High Automation" – ADS takes into account all aspects of driving and driving dynamics and eliminates human control errors caused by "Full Automation" – ADS performs all functions in an automatic mode regardless of the tasks and situation, which requires the development of the entire transport infrastructure and not just the vehicle); (10) modular self-reconfigurable robots that can be grouped from tens to millions of units (homogeneous or several types of heterogeneous ones), change their pairwise connections to form certain structures in solving new problems.

Thus, despite the new tasks of robotics in the field of medicine and transport, the tasks of modular robotics are still far from the final solution and are being actively explored in Russia and abroad. The proposed technological bases for the design of modular robots with pairwise coupling mechanisms describe modern hardware and model-algorithmic tools necessary for the development of promising models of modular robots, each of which is equipped with systems inherent in a fully functional robot, providing its locomotion, communication, navigation, decentralized power and control.

2   Analysis of Modern Modular Robots

The motion of a closed chain-like modular robot in the rolling style is simulated in [8]. The complexity of the problem consists in calculating the position of each block, controlling the effects of the connecting mechanisms between two neighboring blocks, which are necessary for the formation of motion, including on an uneven surface, and for maintaining the stable position of the entire modular robot. The experiments were carried out with a robot UBot during the self-rolling of a closed circuit in the form of a triangle and an oval along different types of surfaces, including the movement of the payload inside the chain.

In the work [9], a new approach to planning the reconfiguration of a modular robot based on graph theory is described. The computational complexity of generating a graph structure is the logarithm of the number of modular robots involved. Calculations were made for modular robots M-TRAN and SuperBot with the number of involved elements of up to 12 pieces.

In the work [10], the problem of redundancy in the design of parallel modular robots is discussed; 13 types of topologies with passive and active elements, used in the problem of changing the shape of the wing, are proposed.

In the work [11], the prospects of modular robotics in the industry are presented. Products and robotic means of production are considered as elements of a cyber-physical system. When designing products, not only a description is formed, but also the composition of modular assembly and processing robots necessary for production.

In the work [12], the problem of the communication between modular robots, a priori not having information about their situation and previous communication sessions, is discussed. In the proposed approach of building a communication network using an infrared channel, neighboring nearby modules are first identified, and then other robots are identified based on the XBee channel. The experiments were carried out with a maximum removal of 10 ModRED modular robots in an enclosed space of up to 30 m and in an open space of up to 90 m.

The review [13] presents the main tasks of rover robotics, behavioral algorithms (clustering, gathering in a pack, collective food search), algorithms for clustering and sorting swarm elements, applied methods of navigation and communication, ways of distributing tasks within a robot swarm. Promising applications, solved on the basis of the methods of swarm robotics, are discussed.

In the work [14], the classification of modular robots is given; trellised and chain-like with 1D-3D topology are distinguished. Linear robots can be serpentine-like and serpenoid. The latter are distinguished by the fact that they move not only due to changes in shape but also to built-in propellers (caterpillars, wheels, etc., installed on the outer surface of each block). For serpentine-like robots, three variants of pairwise connections and changes in the configuration of neighboring blocks are described: (1) in the vertical plane, which ensures the movement of the robot-snake only backwards and forwards; (2) in the horizontal plane, which ensures movement in any direction along the plane; (3) in the vertical and horizontal planes, which ensures the movement of the robot-snake in three-dimensional space. The latter option has a more complex mechanism, consisting of two elements that ensure the rotation of the neighboring block in the vertical and horizontal planes respectively. Also, the work considers 12 principles of the serpentine movement, implemented on the basis of the proposed mathematical models and developed mechanisms of the modular robot snake.

In the work [15], it is proved that the planning of the task of self-reconfiguration of the modular robot is an NP task. To solve it, it is required to use the methods of artificial intelligence. Self-reconfiguration is understood as a reversible process of the movement of interconnected modules without external influence [16]. In self-reconfiguration, problems arise not only in the limited resources of individual modules, but also in collisions in the overall structure, for example: the separation of a modular robot into several parts; the overlap of individual modules in one position in the planning of rebuilding; formation of hollow closed forms, where a part of the modules remains inside and cannot be used further, and vice versa, when some of the modules remain outside and are not available to form an internal structure [17]. When rebuilding the modular robot, it should be taken into account that the modules must remain connected at all steps. This requirement is especially critical for passive modules or blocks that cannot move independently.

The work [18] considers a range of architectures of modular robots depending on their structural parameters. The simplest variant is the group interaction of traditional robots with wheeled/caterpillar propellers, which are interconnected to increase their parameters in solving a particular problem [19]. For robots whose whole body participates in the movement and connection, three types of architectures are distinguished: serpentine-like, lattice, hybrid [20]. In this case, individual modules can take the form of a cube, a cylinder, a parallelepiped, a parallelepiped with rounded corners, trapezoidal, triangular, hexagonal, spherical and others.

For a modular robot consisting of homogeneous blocks with identical functional and physical characteristics, it is somewhat easier, from a computational point of view, to perform self-reconfiguration and solve other tasks. However, the use of universal blocks increases the cost of a modular robot, therefore, in a number of works it is proposed to use heterogeneous blocks, including passive elements. In many modular robots, two types of main and connecting blocks are used [21, 22]. In the work [23], one hundred modular robots of three types with different actuators and sensors were used. In the work [24], 18 types of blocks are used that differ in form, connection mechanisms, methods of movement, capture, the presence of sensors and signal processing facilities.

The weight and size of the blocks used affect the density of the surface formed, which is critical, depending on the application. In the work [25], 34 modular robotic systems with a block mass from 1 to 1000 g and higher were analyzed, and a system of homogeneous modules in combination with passive light components, used to create adaptive furniture, was proposed.

When developing the mechanism for connecting blocks, one must take into account its necessary strength, reversibility and repeatability, power consumption, permissible geometric dimensions, as well as environmental conditions of operation. The mechanical properties of modular couplings impose certain limitations on the space of possible reconfigurations of the modular robot. In most of the works, mechanical, magnetic and chemical methods of connection are used [26]. Recently, methods of connection, based on artificial muscles, have been actively studied [27].

Another important issue in modular robotics is the communication problem, which in most cases is solved with the use of infra-red transceivers [28] or wired communication facilities in the event of a permanent physical connection of all robot modules [29]. Nowadays wireless communication tools based on Bluetooth, ZigBee and other standards of the IEEE 802.15 protocol, as well as IEEE 802.11 WiFi, are most widely used. However, with wireless communication, as the number of modules increases, the problem of signal interference and channel capacity arises, given the small size and limited resources of the modules. In [30], a scalable architecture for multi-channel communication with automatic detection of the nearest transceiver modules and a decentralized communication control system for the operation of the mobile modular robot is proposed.

Thus, the main difference between modular robotics is the availability of separate robotic units that can function independently and be combined into rigidly connected structures to solve more complex spatial problems.

The considered systems solve a wide variety of problems and (theoretically) can realize the control of pairwise connections of homogeneous robots when configuring the swarm into two-dimensional outer contours/shapes located on the surface without making contact connections between themselves. Specificity of the problem considered in this study consists in reconfiguring the three-dimensional arrangement of a swarm of terrestrial homogeneous robots with internal connections to create an unbroken surface of a given shape.

When reconfiguring the swarm, the homogeneous robots must form physical contact junctions with each other and break them at a specific time. Therefore, the task of developing a technical solution for connecting robots to each other is the starting point in modular robotics. The coupling mechanism is one of the main elements in each individual robotic module. The mechanical properties of modular compounds, such as the ultimate strength, reversibility and repeatability of bonds, impose certain limitations on the space of possible reconfigurations of the modular robot. The way the modules communicate with each other can be realized through the physical connection of the modules. In this case, the electrical properties of the connection will impose limitations on the energy consumption of the modules, the size of the batteries, and other characteristics. Also, the connection method directly affects the longevity of the system and the cost of production of each individual module.

3   Onboard Means for Modular Robotics

Currently, SPIIRAS is developing and prototyping the basic components for the creation of specialized computers for on-board computers of robotic complexes and their executive devices: (1) systems on the module (SOM): modules based on digital signal processors (DSP) and ARM-processors; (2) servo controllers with a single control protocol, suitable for use in robots with a large number of degrees of freedom; (3) controllers for the integration of robot subsystems with a modular architecture.

System modules form the basis for creating embedded computers. They comprise the main system components that have a high degree of integration: processor, RAM, non-volatile memory, power controllers, coprocessors. Due to this, the system modules can be re-used, which allows increasing their circulation and reduce the cost of the final computers and robotics constructed with their use [31–34].

Figure 1 shows the developed SMARC module based on the ARM processor – SMARC-AM335x. The module includes an ARM Cortex A8 processor with a frequency of 600 to 1000 MHz, DDR3 memory of 256 to 1024 MB, non-volatile memory of 4 GB. The advantage of the module is low cost (about $ 25), small dimensions (82 * 50 * 5 mm) and low power consumption (2–3 W). The module allows creating not only small on-board computers for robots but also user interface systems equipped with displays and touch screens. The module is made according to the SMARC standard, one of the most advanced international standards for system modules.

Fig. 1. The developed module SMARC-AM335x

To increase the capabilities and reduce the cost of robots, the hardware-software of the model servo was performed. A schematic diagram was created and the circuit board of the controller was made; software was developed for updating and controlling the controller; a working prototype of the product was assembled. This made it possible to receive the information about the angle and other parameters of the servo drive, for example, current consumption, voltage, speed, and also to change the regulation law – for example, PID controller parameters depending on the task. The appearance of the developed controller and the modernized servo drive is shown in Fig. 2.

The software-hardware of the model servo developed during the modernization allows creating a series of controlled drives for use in robotic complexes. The proposed

Fig. 2. The developed controller and upgraded RC-servo

solution improves the dynamic characteristics of the servo, enables the creation of a network of controlled drives, and also reduces the cost of the servo and the robot in comparison with existing analogues.

Controllers for the integration of subsystems are similar to devices like USB Hub. Their main task is to ensure connectivity of information transmission between the central on-board computer and the final controllers of servo drives and sensors. The problem with the USB bus is a limited number of devices on one bus (up to 127), as well as bus idle time between requests of the central node. Subsystem integration controllers allow individual bus segments to operate autonomously from the central node, so that limited bus bandwidth can be distributed between a significantly larger number of devices. Also, the application task of integration controllers consists in the rapid branching of the control network.

4 Conclusion

The analysis of projects on modular robotics in the publications of the last ten years has confirmed the relevance of this scientific direction and revealed the prospects of attracting cloud robotics techniques for distributed sensory data collection, management and centralized training using stationary computing resources.

Developed on-board computers based on SMARC-modules are planned to be used for the processing of audiovisual signals, the remote control of modular robotic complexes, as well as in embedded facilities of intelligent cyberphysical space for designing systems with a multimodal interface, which realize natural interaction between users and automatic services, including for contactless determination of biometric indicators based on speech signal processing and analysis of other nonverbal modalities of human behavior.

Acknowledgments. This work is partially supported by the Russian Foundation for Basic Research (grant № 16-29-04101-ofi_m).

References

1. Fukuda, T., Nakagawa, S.: Approach to dynamically reconfigurable robotic system. In: IEEE International Conference on Robotics and Automation, ICRA, pp. 1581–1586 (1988)
2. Ahmadzadeh, H., Masehian, E.: Modular robotic systems: methods and algorithms for abstraction, planning, control, and synchronization. Artif. Intell. **223**, 27–64 (2015)
3. Rubenstein, M., Ahler, C., Hoff, N., Cabrera, A., Nagpal, R.: Kilobot: a low cost robot with scalable operations designed for collective behaviors. Robot. Auton. Syst. **62**, 966–975 (2014)
4. McKee, G.: What is networked robotics? Inform. Control Autom. Robot. **15**, 35–45 (2008)
5. Kuffner, J.: Cloud-enabled robots. In: IEEE-RAS International Conference on Humanoid Robots (2010)
6. Michniewicz, J., Reinhart, G.: Cyber-physical robotics – automated analysis, programming and configuration of robot based on cyber-physical-systems. Procedia Technol. **15**, 566–575 (2014)
7. Kopacek, P.: Development trends in robotics. IFAC-PapersOnLine **49**(29), 36–41 (2016)
8. Wang, X., Jin, H., Zhu, Y., Chen, B., Bie, D., Zhang, Y., Zhao, J.: Serpenoid polygonal rolling for chain-type modular robots: A study of modeling, pattern switching and application. Robot. Comput.-Integr. Manuf. **39**, 56–67 (2016)
9. Taheri, K., Moradi, H., Asadpour, M., Parhami, P.: MVGS: a new graphical signature for self-reconfiguration planning of modular robots based on multiple views theory. Robot. Auton. Syst. **79**, 72–86 (2016)
10. Moosavian, A., Xi, F.: Modular design of parallel robots with static redundancy. Mech. Mach. Theory **96**, 26–37 (2016)
11. Michniewicz, J., Reinhart, G.: Cyber-physical-robotics – modeling of modular robots for automated planning and execution of assembly tasks. Mechatronics **34**, 170–180 (2016)
12. Baca, J., Woosley, B., Dasgupta, P., Nelson, C.: Configuration discovery of modular self-reconfigurable robots: real-time, distributed, IR+XBee communication method. Robot. Auton. Syst. **91**, 284–298 (2017)
13. Bayındır, L.: A review of swarm robotics tasks. Neurocomputing **172**(8), 292–321 (2016)
14. Gonzalez-Gomez, J., Zhang, H., Boemo, E.: Locomotion principles of 1D topology pitch and pitch-yaw-connecting modular robots. In: Habib, M.K. (ed.) Bioinspiration and Robotics: Walking and Climbing Robots, pp. 403–428. InTech (2007)
15. Gorbenko, A., Popov, V.: Programming for modular reconfigurable robots. Program. Comput. Softw **38**(1), 13–23 (2012)
16. Grob, R., Bonani, M., Mondada, F., Dorigo, M.: Autonomous self-assembly in swarm-bots. IEEE Trans. Robot. **22**(6), 1115–1130 (2006)
17. Stoy, K., Brandt, D., Christensen, D.J.: Self-Reconfigurable Robots: An Introduction. The MIT Press, Cambridge (2010)
18. Moubarak, P., Ben-Tzvi, P.: Modular and reconfigurable mobile robotics. Robot. Auton. Syst. **60**, 1648–1663 (2012)
19. Wang, W., Yu, W., Zhang, H.: JL-2: a mobile multi-robot system with docking and manipulating capabilities. Int. J. Adv. Robot. Syst. **7**(1), 9–18 (2010)
20. Karagozler, M.E., Thaker, A., Goldstein, S.C., Ricketts, D.S.: Electrostatic actuation and control of micro robots using a post-processed high-voltage SOI CMOS chip. In: Proceedings of the IEEE International Symposium on Circuits and Systems, ISCAS 2011, pp. 2509–2512 (2011)

21. Baca, J., Hossain, S.G.M., Dasgupta, P., Nelson, C.A., Dutta, A.: ModRED: hardware design and reconfiguration planning for a high dexterity modular self-reconfigurable robot for extra-terrestrial exploration. Robot. Auton. Syst. **62**, 1002–1015 (2014)
22. Fan, J., Zhang, Y., Jin, H., Wang, X., Bie, D., Zhao, J., Zhu, Y.: Chaotic CPG based locomotion control for modular self-reconfigurable robot. J. Bionic Eng. **13**, 30–38 (2016)
23. Levi, P., Meister, E., Schlachter, F.: Reconfigurable swarm robots produce self-assembling and self-repairing organisms. Robot. Auton. Syst. **62**, 1371–1376 (2014)
24. Moses, M.S., Ma, H., Wolfe, K.C., Chirikjian, G.S.: An architecture for universal construction via modular robotic components. Robot. Auton. Syst. **62**, 945–965 (2014)
25. Sprowitz, A., Moeckel, R., Vespignani, M., Bonardi, S., Ijspeert, A.J.: Roombots: a hardware perspective on 3D self-reconfiguration and locomotion with a homogeneous modular robot. Robot. Auton. Syst. **62**, 1016–1033 (2014)
26. Neubert, J., Rost, A., Lipson, H.: Self-soldering connectors for modular robots. IEEE Trans. Robot. **30**(6), 1344–1357 (2014)
27. Mathijssen, G., Schultz, J., Vanderborght, B., Bicchi, A.: A muscle-like recruitment actuator with modular redundant actuation units for soft robotics. Robot. Auton. Syst. **74**, 40–50 (2015)
28. Sproewitz, A., Laprade, P., Bonardi, S., Mayer, M., Mockel, R., Mudry, A., Ijspeert, A.: Roombots-towards decentralized reconfiguration with self-reconfiguration modular robotic metamodules. In: Proceedings of IEEE/RSJ IROS, pp. 1126–1132 (2010)
29. Garcia, R., Lyder, A., Christensen, D., Stoy, K.: Reusable electronics and adaptable communication as implemented in the odin modular robot. In: Proceedings of IEEE ICRA, pp. 1152–1158 (2009)
30. Kuo, V., Fitch, R.: Scalable, multi-radio communication in modular robots. Robot. Auton. Syst. **62**, 1034–1046 (2014)
31. Shlyakhov, N.E., Vatamaniuk, I.V., Ronzhin, A.L.: Survey of methods and algorithms of robot swarm aggregation. IOP Conf. Ser. J. Phys. **803**, 012146 (2017)
32. Kodyakov, A.S., Pavlyuk, N.A., Budkov, V.Y., Prakapovich, R.A.: Stability study of anthropomorphic robot antares under external load action. IOP Conf. Ser. J. Phys. **803**, 012074 (2017)
33. Burakov, M.V., Shyshlakov, V.F.: A modified smith predictor for the control of systems with time-varying delay. SPIIRAS Proc. **51**(2), 60–77 (2017)
34. Gaponov, V.S., Dashevsky, V.P., Bizin, M.M.: Upgrading the firmware of model servos for use in anthropomorphic robotic systems. Dokl. TUSUR **19**(2), 41–50 (2016)

Context-Awared Models in Time-Continuous Multidimensional Affect Recognition

Dmitrii Fedotov[✉], Maxim Sidorov, and Wolfgang Minker

Institute of Communications Engineering, Ulm University,
Albert Einstein-Allee, 43, 89081 Ulm, Germany
{dmitrii.fedotov,wolfgang.minker}@uni-ulm.de,
maxim.sidorov@alumni.uni-ulm.de

Abstract. Modern research in the field of automatic emotion recognition systems are often dealing with acted affective databases. However, such data is far from the real-world problems. Human-computer interaction systems are extending their field of application and becoming a great part of human's everyday life. Such systems are communicating with user through dialog and are supposed to define the current mood in order to adjust their behaviour. To increase the depth of emotion definition, multidimensional time-continuous labelling is used in this study instead of utterance-label categorical approach. Context-aware (long short-term memory recurrent neural network) and context-unaware (linear regression) models are contrasted and compared. Different meta-modelling techniques are applied to provide final labels for each dimension based on unimodal predictions. This study shows that context-awareness can lead to a significant increase in emotion recognition precision with time-continuous data.

Keywords: Context-learning · Long short-term memory · Time-continuous data · Multi-dimensional data

1 Introduction

Human-human interaction in real life has two main aspects: information contained in speech and emotion expressed by the human. Speech recognition techniques allow computers to understand human speech but they lack an emotional component. Exactly the same words of phrases can have completely different meanings if said with different emotions. It can be a statement, a question or a guess.

Computer-based systems have become a great part of our everyday lives. However, it is necessary for computers to understand human emotions in order to succeed in interaction with them. This became obvious with the emergence of personal assistants such as Siri, Cortana, Google Now and others. These systems try to interact with humans in a more natural way, understanding and answering non-standardized questions, similar to the interaction between two humans. One of the most important parts of human understanding is the ability to identify and react to emotions. Emotion recognition may significantly improve the quality of human-robot interaction, speech recognition systems and artificial intelligence in general.

© Springer International Publishing AG 2017
A. Ronzhin et al. (Eds.): ICR 2017, LNAI 10459, pp. 59–66, 2017.
DOI: 10.1007/978-3-319-66471-2_7

Most of the research on emotion recognition deals with utterance-level categorical data labeling, i.e. each data sample had one label from the list, e.g. anger, happiness, neutral. However, recent research has focused on the dimensional data that provides more flexibility and precision of emotion definition. Besides the multidimensional labelling, new, more complex modalities are included in modern research along with audio-visual data. One of the examples of such modalities are physiological data, e.g. electrocardiogram (ECG) and electro-dermal activity (EDA) [1].

A combination of different modalities and multi-dimensional labelling allows meta-models to be built for emotion recognition systems that are capable of providing precise predictions. As shown in this paper, models based on contextual data have achieved an increase of up to 89% in the performance measured by concordance correlation coefficient (CCC) [2] in comparison to that shown by context-unaware models.

This paper is structured as follows: Sect. 2 provides an overview of research related to multimodal, multi-dimensional emotion recognition; Sect. 3 details the data used in this study and the preprocessing procedures; in Sect. 4 the methods used are described; in Sect. 5 experimental results are shown and analysed; conclusions from this study and proposed future research are presented in Sect. 6; acknowledgements for this study are presented below the conclusions.

2   Related Work

Understanding emotions plays a significant role in human social life. Emotion recognition is developed in childhood and helps the child to successfully interact with other humans. Children with better understanding of human emotions have greater chances of building strong, positive relationships in their future lives [3]. In everyday life, humans express emotion through various channels, e.g. voice, mimic, gestures, eyes movement, etc. Although humans are the best emotion recognition systems ever built they can understand the same expression differently due to the subjective nature of emotion. Aside from cultural and social differences, the ability to recognize emotions can vary greatly with age [4].

There are many different emotions, and therefore one of the problems while building an emotion recognition system is to extract the most common and important among them. For example, authors in [5] trained their system to recognize the following emotions: joy, teasing, fear, sadness, disgust, anger, surprise and neutral. The authors used a neural network as classifier and tried different strategies: One-Class-in-One Neural network (OCON) [6], All-Class-in-One Neural Network (ACON) and a single-layer neural network using the Learning Vector Quantisation (LVQ) method [7].

One of the ways to avoid the problem of emotion categorization is to use a multidimensional approach for emotion definition. Several different scales may be used to describe a particular emotion, but problems are often reduced to using two scales: valence and arousal [8].

Another study was focused on building a neural network that can handle different modalities (voice, text and facial expressions) [9]. The features for emotion recognition systems are not limited to the ones extracted from video, audio and text data. The authors in [10] used an Electroencephalogram (EEG) signals database called DEAP [11] as input data and a multi-class Support Vector Machine as a classifier.

Some of the recent research were focused on emotion recognition under uncontrolled conditions in wild using audio-visual data [12]. Authors used basic and kernel extreme learning machines as classifiers. However, modern research tends to work with spontaneous, time-continuous data. Recurrent neural networks have been used in recent studies for continuous predictions due to their ability to model sequences and use information from the past [13].

3 Data Preprocessing

In this study, a multimodal multi-dimensional corpus of spontaneous human-human interaction called RECOLA (Remote COLlaborative and Affective interactions) was used [14]. The database consists of spontaneous interactions in French, collected during the solving of a cooperative problem. There are 23 participants aged between 18 and 25 years old and having different mother tongues: French, German and Italian.

Information during interaction was collected in 4 modalities: audio, video, ECG and EDA. Each recording has a duration of 5 min and time-continuous two-dimensional labels (arousal and valence) from 6 French-speaking annotators. Physiological information, i.e. ECG and EDA, is missing for 5 participants, thus the corresponding speakers were not included in the research of this study.

The database contains raw data as well as extracted features for each modality. Due to technical issues, different modalities have different amounts of information included in the feature set, e.g. the physiological feature set lacks the data for the first and last two seconds of each recording. Only an intersection of data was used in this study to perform fully multimodal emotion recognition.

The audio features set contains 65 low-level descriptors (LLD) e.g. spectral, energy and voicing related, with their first order derivate giving us 130 audio features in total. The video features set contains 20 LLDs (15 facial action units, 3D head pose, mean and standard deviation of the optical flow in the region around the head) and their first order derivate (40 features). ECG feature set includes 28 LLDs based on heart rate and spectral data with their first order derivate for all LLDs except two of them (54 features). EDA feature set contains 30 LLDs based on skin conductance response, skin conductance level, normalized length density and non-stationary index with their first order derivate (60 features).

For each combination of train, development and test sets, features were normalized to zero mean and unit variance based on the parameters from the training set. Assessments of emotion obtained by annotators were averaged frame-wise for each dimension. To perform context-aware time-continuous emotion recognition, the data was preprocessed to meet the requirements of a recurrent neural network. Although there is no clear information about the optimal time-window for emotion recognition [15] in this study exactly one second was fed to the neural network. The original features were extracted with a frequency of 25 Hz and after the preprocessing, each sample of the set has information about the current and 24 previous frames. To reduce the instability of the output curve, a smoothing procedure was applied to the network output. The network was constructed in order to make a prediction for the current and subsequent 24 frames. The prediction for a particular frame obtained on different time steps was averaged to provide more smooth and reliable results. When the information on the previous or subsequent frames was unavailable, e.g. for the first frames in feature set and the last frames in a label set, the gaps were filled with zeros to avoid any loss of information.

4 Methodology

In this study, context-unaware methods were compared to ones that are suitable for performing time-continuous contextual prediction. Context-awareness allows the information about previous and/or past events to be used to make a prediction of the system state at the current moment. A linear regression model with L2 regularization was chosen as a baseline, context-unaware model. It had a regularization strength of 0.1 and precision of the solution equal to 10^{-5}.

As a context-aware model, a long short-term memory (LSTM) recurrent neural network (RNN) was used [16]. RNNs take information about the previous time step through feedback loop while defining the current system state. Traditional RNNs suffer from vanishing gradient problem when errors and activation of the recurrent connections decay exponentially. This problem leads to a limitation of context that can be taken into consideration by neural network. LSTM RNNs can avoid this problem with fine regulation of system state by special gates: input gate, output gate and forget gate. These gates allow to accumulate the information about previous time steps over long duration and drop the information when needed. The weights of self-loops are not fixed, but based on the gates which allows to change the level of data integration.

The LSTM-RNN had two hidden layers with 80 and 60 LSTM blocks respectively as this architecture has shown good performance previously [14]. The size of the input layer was defined by the input vector of corresponding modality and the output vector had 25 neurons for the subsequent frame prediction. To avoid overfitting, recurrent layers were followed by the dropout layers [17] with the probability of connections being dropped equal to 0.1. The LSTM blocks had a ReLU activation function [18] and the neurons of the output layer had a simple linear activation function. The LSTM models were optimized by root mean square propagation (RMSprop) using the mean squared error as a loss function. LSTM implementation is provided by Keras [19], other models are provided by scikit-learn libraries for Python [20].

The described models were used to obtain unimodal predictions for each emotion dimension, which were further fed to meta-models that defined the final labels for each time step of the data. In the scope of this study, 4 models were considered: (i) linear regression (LR) with regularization strength = 0.1; (ii) multilayer perceptron (MLP) with three optimisers: adam [21], standard gradient decent (SGD), lbfgs [22] (learning rate = 10^{-5}); (iii) decision trees (DT) [23]; (iv) random forest (RF) [24]. The parameters not mentioned above were kept at their default values.

The pipeline of emotion recognition system used in this study is presented at Fig. 1. Dashed arrows represent contextual data, e.g. each sample of dataset contains not only the current frame, but also previous (features) or further (labels) frames.

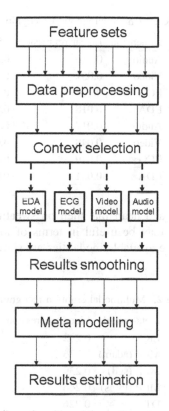

Fig. 1. Pipeline of multimodal emotion recognition system

5 Experimental Results

To compare the proposed models, a speaker-independent cross-validation was performed. Unimodal models were trained on the train set, which consists of N-3 speakers (where N is the number of speakers). The meta-model was trained on the development set containing 2 speakers and the final results were computed on the test set with one speaker left. This procedure was repeated until each speaker was included in one of the test sets. In this paper, our comparison is based on the CCC metric which is a combination of the mean squared error and Pearson's correlation coefficient. The result measured with CCC may reach the maximum value of 1 meaning the absolute identity of label curves. The proposed methods were compared at the unimodal and multimodal levels. The results of the unimodal prediction are shown in Table 1. Bold values represent the best results achieved in defined condition.

Context-awareness allows the emotion recognition system to provide more precise predictions for all modalities, except EDA. The performance of models on different modalities differs greatly and one may conclude that it will affect the contribution of the modality at decision level fusion. For example, audio and ECG data for arousal dimension will play a much more significant role than EDA as the intensity of emotion

Table 1. Unimodal emotion recognition

Modality	Dimension	Linear regression	LSTM
Arousal	Audio	0.116	**0.375**
	Video	0.051	**0.093**
	ECG	0.124	**0.192**
	EDA	**0.018**	0.016
Valence	Audio	0.017	**0.103**
	Video	0.048	**0.097**
	ECG	0.049	**0.050**
	EDA	**0.051**	0.006

has greater impact on voice and heart rate then on sweating. However, information derived from each modality may be useful in terms of meta-model. The results of multimodal fusion over all the available modalities are shown in Table 2.

Table 2. Multimodal emotion recognition

Dimension	Meta-model	Linear regression	LSTM
Arousal	LR	0.232	**0.476**
	MLP (adam)	0.236	0.409
	MLP (sgd)	0.239	0.370
	MLP (lbfgs)	0.246	0.397
	DT	0.226	0.330
	RF	0.249	0.377
Valence	LR	0.151	0.202
	MLP (adam)	0.153	0.280
	MLP (sgd)	0.139	0.115
	MLP (lbfgs)	0.145	**0.288**
	DT	0.134	0.228
	RF	0.132	0.267

The performance of the emotion recognition system for both modalities had significantly improved with the use of recurrent neural networks. The contextual predictions obtained at the unimodal stage provide a much better basis for multimodal data fusion.

6 Conclusions and Future Work

Multidimensional time-continuous emotion recognition has a wide range of applications and is closer to human emotion perception than utterance-level predictions. Context-awareness and multimodal decision-level data fusion led to a great increase in performance along with computational costs. To reduce the time for models training, data sparsing procedures may be applied and its effect on the system performance ought to be analysed as well as effect of different time window sizes for data preprocessing and prediction smoothing in order to find optimal values of the useful frames for each modality.

The contribution of different modalities during the fusion can be investigated and used to build a more precise multimodal system, e.g. some of the modalities can be dropped for a particular dimension or the feature-based fusion of a pair of modalities may be used at the contextual prediction stage.

Acknowledgments. The work presented in this paper was partially supported by the Transregional Collaborative Research Centre SFB/TRR 62 "Companion-Technology for Cognitive Technical Systems" which is funded by the German Research Foundation (DFG).

References

1. Ringeval, F., Sonderegger, A., Sauer, J., Lalanne, D.: Introducing the RECOLA multimodal corpus of remote collaborative and affective interactions. In: Proceedings of IEEE on Face and Gestures 2013, 2nd International Workshop on Emotion Representation, Analysis and Synthesis in Continuous Time and Space (EmoSPACE), Shanghai, China (2013)
2. Lawrence, I., Kuei, L.: A concordance correlation coefficient to evaluate reproducibility. Biometrics **45**, 255–268 (1989)
3. Denham, S.A.: Emotional Development in Young Children. Guilford Press, New York (1998)
4. Chronaki, G., et al.: The development of emotion recognition from facial expressions and non-linguistic vocalizations during childhood. Br. J. Dev. Psychol. **33**(2), 218–236 (2015)
5. Nicholson, J., Kaxuhiko, T, Nakatsu, R.: Emotion recognition in speech using neural networks. In: Proceedings of 6th International Conference on Neural Information Processing (ICONIP 1999), vol. 2. IEEE (1999)
6. Markel, J.D., Gray, A.H.: Linear Prediction of Speech. Springer, New York (1976)
7. Kohonen, T.: Self-Organizing Maps. Springer Series in Information Sciences, vol. 30. Springer, Heidelberg (1995)
8. Russell, J.A., Bachorowski, J.A., Fernández-Dols, J.M.: Facial and vocal expressions of emotion. Ann. Rev. Psychol. **54**(1), 329–349 (2003)
9. Fragopanagos, N., Taylor, J.G.: Emotion recognition in human-computer interaction. Neural Netw. **18**(4), 389–405 (2005)
10. Vijayan, A.E., Deepak, S., Sudheer, A.P.: EEG-based emotion recognition using statistical measures and auto-regressive modeling. In: IEEE International Conference on Computational Intelligence and Communication Technology (CICT). IEEE (2015)
11. Koelstra, S., et al.: DEAP: a database for emotion analysis; using physiological signals. IEEE Trans. Affect. Comput. **3**(1), 18–31 (2012)
12. Kaya, H., Salah, A.: Combining modality-specific extreme learning machines for emotion recognition in the wild. J. Multimodal User Interfaces **10**(2), 139–149 (2016)
13. Wöllmer, M., et al.: LSTM-modeling of continuous emotions in an audiovisual affect recognition framework. Image Vis. Comput. **31**(2), 153–163 (2013)
14. Ringeval, F., Eyben, F., Kroupi, E., Yuce, A., Thiran, J.-P., Ebrahimi, T., Lalanne, D., Schuller, B.: Prediction of asynchronous dimensional emotion ratings from audio-visual and physiological data. Pattern Recogn. Lett. **66**, 22–30 (2015)

15. Gunes, H., Pantic, M.: Automatic dimensional and continuous emotion recognition. Int. J. Synth. Emot. **1**(1), 68–99 (2010)
16. Hochreiter, S., Schmidhuber, J.: Long short-term memory. Neural Comput. **9**(8), 1735–1780 (1997)
17. Srivastava, N., et al.: Dropout: a simple way to prevent neural networks from overfitting. J. Mach. Learn. Res. **15**(1), 1929–1958 (2014)
18. Nair, V., Hinton, G.E.: Rectified linear units improve restricted Boltzmann machines. In: Proceedings of the 27th International Conference on Machine Learning (ICML 2010) (2010)
19. Francois, C.: Keras (2015). https://github.com/fchollet/keras
20. Pedregosa, F., et al.: Scikit-learn: machine learning in python. JMLR **12**, 2825–2830 (2011)
21. Kingma, D., Ba, J.: Adam: a method for stochastic optimization (2014). arXiv preprint arXiv:1412.6980
22. Liu, D.C., Nocedal, J.: On the limited memory BFGS method for large scale optimization. Math. Program. **45**(1), 503–528 (1989)
23. Breiman, L., et al.: Classification and regression trees. Wadsworth & Brooks, Monterey (1984)
24. Breiman, L.: Random forests. Mach. Learn. **45**(1), 5–32 (2001)

Facing Face Recognition with ResNet: Round One

Ivan Gruber[1,2,3]($\boxtimes$), Miroslav Hlaváč[1,2,3], Miloš Železný[1], and Alexey Karpov[3,4]

[1] Faculty of Applied Sciences, Department of Cybernetics,
UWB, Pilsen, Czech Republic
grubiv@ntis.zcu.cz, {mhlavac,zelezny}@kky.zcu.cz
[2] Faculty of Applied Sciences, NTIS, UWB, Pilsen, Czech Republic
[3] ITMO University, St. Petersburg, Russia
karpov@iias.spb.su
[4] SPIIRAS, St. Petersburg, Russia

Abstract. This paper presents initial experiments of an application of deep residual network to face recognition task. We utilize 50-layer deep neural network ResNet architecture, which was presented last year on CVPR2016. The neural network was modified and then fine-tuned for face recognition purposes. The method was trained and tested on challenging Casia-WebFace database and the results were benchmarked with a simple convolutional neural network. Our experiments of classification of closed and open subset show the great potential of residual learning for face recognition.

Keywords: Face recognition · Classification · Neural networks · Data augmentation · Residual training · Computer vision

1 Introduction

Face recognition has been one of the most intensively studied topics in computer vision for last few decades and received great attention because of its applications in various real-world problems, among which can be counted security and surveillance systems, general identity recognition or person database investigation. The most significant usage of face recognition is in biometrics. Compared with some other biometrics techniques (fingerprints, iris, etc.) face recognition has the potential to non-intrusively recognize subject without any further cooperation of the subject. This and the fact, that camera sensors are much cheaper than various fingerprints or iris scanners, make face recognition most attractive biometric technique. Another important thing about face recognition is, that humans identify other people according to their face too, therefore they are likely to be comfortable with systems that use this approach.

In this article, we presented a method for face classification in unconstrained conditions based on an residual neural network [1]. The method is trained and

© Springer International Publishing AG 2017
A. Ronzhin et al. (Eds.): ICR 2017, LNAI 10459, pp. 67–74, 2017.
DOI: 10.1007/978-3-319-66471-2_8

also tested on the major part of Casia-WebFace database [2]. We performed two types of experiments: (1) face classification of the closed subset; (2) face classification of the open subset; both with very promising results.

This paper is organized as follows: in Sect. 2 we provide quick look on the literature in this area; in Sect. 3 we describe used database and our data augmentations; in Sect. 4 we discuss presented method; in Sect. 5 we show obtained experimental results, while we draw conclusions and discuss future research in Sect. 6.

2   Related Work

In recent years, face recognition is experiencing renaissance through the deep convolutional neural networks (CNNs). After the breakthrough article from Krizhevsky et al. [3] in 2012, many novels approaches using CNNs for face recognition appeared. To name the most important ones, let's mention following works. In 2014, Taigman et al. [4] with their method based on CNN and effective face alignment significantly improved recognition rate on the LFW dataset [5], standard benchmark dataset for face recognition. Later in 2014, Sun et al. [6] presented method using CNN to learn effective features for face recognition. They further improved their features in article [7]. In 2015, Lu and Tang [8] surpassed human-level face verification performance on LFW using novel method based on GaussianFaces. Finally, Schroff et al. [9] reached almost flawless performance on LFW with their deep neural network trained to provide 128-D feature vector using triplet-based loss. It should be noted, that their method doesn't need any face alignment. This is caused by enormous amount of used training data (260 million images).

Despite all of the above-mentioned methods, the obtained state-of-the-art results are unrealistically optimistic from the position of pose-invariant face recognition, because most of the LFW images can be classified as near-frontal. Therefore most of existing algorithms can have problems with pose variances. To address this problem there were created several datasets (for example CASIA WebFace [2] or IJB-A [10]) containing variable external conditions including extreme pose variances. In 2016, Masi et al. [11] improved state-of-the-art results for datasets including large pose variances by presenting method based on Pose-Aware CNN Models.

Later in 2016, He et al. [1] proposed novel DNN architecture containing residual learning blocks to address a problem of degradation during learning very deep networks. This entry, which won the 1st place in ILSVRC-2015 [12], has very high potential for face recognition and it is probably the biggest upgrade of neural networks since Krizhevsky's AlexNet.

3   Dataset

As a training set for tested neural networks was chosen Casia-WebFace database. Casia-WebFace database is the second biggest publicly available database for face recognition (the biggest one is The Megaface dataset [13]) and contains 494414 RGB images of 10575 subjects with resolution of 250 × 250 pixels. All images were semi-automatically collected from the Internet, i.e. persons are captured in variable conditions including pose, illumination, occlusion, age variations, haircut changes, sunglasses, etc. Most faces are centered on the images. For exemplary images see Fig. 1.

Fig. 1. Exemplary images from Casia-WebFace database

For the training, we decided to use only identities, which have at least 100 images presented. This leaves us with 181901 images for 925 identities. In order of CNN training, the resolution of all images was decreased to 64 × 64 pixels. To enrich these data we generated flipped version of each image. To further enriching training set and balancing counts of images per identity (which is very important for neural network training) we used data augmentation. To be more concrete, we modified images with Gaussian blur, noise, and brightness transformations. This leads to 908953 images in total. The database was appropriately split (each unique image with all its augmentations was in the same subset) into three subsets - training, validation and testing set, whereas 70% of images were used for the training set, 15% for the validation (development) set and 15% for the testing set.

4   Method

Due to the neural networks improvements in recent years, most hand-crafted feature descriptors for face recognition, if enough data is available, become obsolete. This results from the fact, that such descriptors use the same, hardly optimal, operator to all locations of the face. On the other hand, learning-based descriptors can discover optimal operators for each location in the facial image.

Firstly, we train simple CNN on a multi-class (925 classes) face recognition task by minimizing cross-entropy loss to obtain benchmark results. The input is an RGB face image with size 64×64 pixels. The CNN's architecture contains three convolutional layers (32 filters each), each followed by ReLU non-linearity and max-pooling layer (to fight with overfitting and more robustness to local translation). The architecture is ended with a fully-connected layer containing 925 neurons (one for each class). For updating CNN's parameters W we used standard SGD optimization method. CNN was trained with the mini-batch size of 64 images during 850k iterations.

Secondly, we train deep residual network (DRN) with an architecture based on ResNet-50. DRN was proposed by He et al. [1] to address a problem of degradation during learning very deep neural networks. During testing, authors observed counter-intuitive phenomenon - adding more layers to the architecture causes higher training error. Historically this problem occurred, because of vanishing/exploding gradients during backpropagation, however, this phenomenon has been largely addressed by normalized initialization and intermediate normalization layers. Therefore, this degradation of training accuracies indicates that not all systems are similarly easy to optimize.

Authors address the degradation problem by introducing a deep residual learning framework. Instead of learning each group of stacked layers directly fit a desired underlying mapping, they let these layers fit a residual mapping. Let $H(x)$ be the desired underlying mapping of a group of stacked layers with x denoting the input to the first of these layers. Based on the hypothesis that multiple nonlinear layers can asymptotically approximate complicated functions, multiple nonlinear layers should be able to asymptotically approximate the residual function, i.e. $F(x) = H(x) - x$. The original function thus becomes $F(x) + x$. Although both forms should be able to asymptotically approximate the desired functions, the ease of learning is different. The formulation of $F(x) + x$ we realized by the shortcut connections as element-wise addition (Fig. 2).

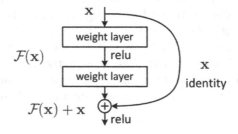

Fig. 2. Residual learning: a building block [1]

Formally building block is defined as:

$$y = F(x, \{W_i\}) + x, \tag{1}$$

where x is the input vector, y is the output vector of the layers and $F(x, \{W_i\})$ represents the residual mapping to be learned. The main advantage of this configuration is that the computational complexity of element-wise addition is negligible. The dimensions of x and F must be equal, however, if this is not the case, then linear projection of x can be performed. The function F can represent multiple fully-connected or convolution layers, in the later case the element-wise addition is performed channel by channel. He et al. experimented with F that contains one, two or three layers, however, if F has only a single layer, no advantages were observed.

As was already said, we trained NN with the architecture based on their 50-layer residual network. The network contains in total 16 residual learning building blocks each followed by ReLU non-linearity. Function F represents three convolutional layers, first two are also followed by ReLU non-linearity, in all cases. The only two differences between our and their architecture are the different size of the last (fully-connected) layer and the different size of the last average-pooling layer in ours. This second change was made because our input to the network has size 64×64 instead of original 224×224 pixels. The initial idea was to trained ResNet from the beginning with random initialization of weights W, however, we believe, that low-level features extracted in the original classification task (ILSVRC-2015) are for face recognition relevant enough, that we can utilize already pre-trained weights of the original ResNet-50. Therefore, we decided to perform fine-tuning of original weights, whereas weights of convolutional layers in first 14 residual learning blocks were fixed. ResNet-50 was trained with the batch size of 6 during 120k iterations. For updating weight parameters W we used SGD optimization again.

Both neural networks were implemented in Python using Caffe deep learning framework [14].

5 Experiments and Results

In this section, we present the experimental results of our two experiments. Both experiments were performed on our test set described in Sect. 3.

5.1 Classification of the Closed Subset

In our first experiment, we evaluate our networks on the face classification of closed subset task. This type of task is easier than the classification of the open subset, because we don't allow any face image of foreign identity to come to the input of the network, therefore we don't need any thrash class or confidence threshold. NN simply always classifies the input face image into one of 925 trained classes.

Simple CNN reached after 850k training iterations only average results, to be more concrete 72.4% and 71.0% recognition rate (RR) on the development set and the test set respectively. It should be noted, that it can not be expected

Table 1. Comparison of classification recognition rates

Method	Development set	Test set
CNN	72.4%	71.0%
ResNet-50-60	89.3%	87.9%
ResNet-50-120	91.5%	90.7%

to obtain state-of-the-art results with such simple architecture. Nevertheless, we trained this network primarily to get benchmark results.

First, we fine-tune ResNet-50 with only 60k iterations. Thus trained network already reaches very promising result (89.3% and 87.9% RR). Next, we double the number of training iterations to reach even better results - 91.5% and 90.7% RR on development and test set respectively, which are very good rates on such challenging database as Casia-WebFace. We do not reach better results with further increasing of the number of iterations. Nevertheless, we assume, that with bigger training set, we can reach even better results and we definitely will examine this possibility in our future research. The comparison of results of classification is showed in Table 1.

Overall, this experiment showed the great potential of ResNet and residual learning for face recognition.

5.2   Classification of the Open Subset

In our second experiment, we evaluate our method on the face classification of open subset task. As it was already suggested, classification of closed subset can be simply extended by implementing confidence threshold. If the value of probability for the most probable class would be below this threshold, person's identity would be claimed as an unknown. It would be classified without any change otherwise. As the unknown persons, we use facial images of identities discarded during creating training set (all identities with less than 100 images presented). Moreover, we utilize facial images from the MUCT face database [15].

The optimal confidence threshold is found on the basis of false acceptance rate and false rejection rate, see Fig. 3. As you can see, best possible results are reached by choosing threshold value approximately equal to 0.974, however, it should be noted, that for practical usage both rates are not equally problematic. Usually, it is not a problem to repeat login process into your bank account, because of false rejection of the system. But if someone else logs into your bank account, it will be definitely considered as a big security hole in the system. Therefore, we believe, that more strict threshold, 0.984 for example, is actually better.

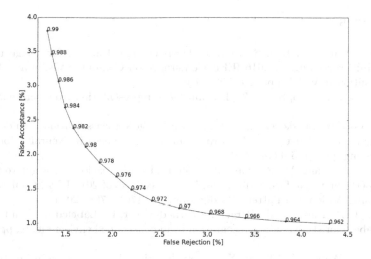

Fig. 3. False acceptance vs false rejection for open subset classification

6 Conclusion and Future Work

Face Recognition is a very challenging problem, which, despite all the effort and popularity in recent years, can be in uncontrolled conditions hardly solved. Our work shows that residual training of deep neural networks has big potential for face recognition tasks. We believe, that with some minor modifications of ResNet architecture and more augmentations of the training set, or with bigger training set, has residual training of neural networks potential to improve state-of-the-art results.

In our future research, we would like to focus on the testing of ResNet on verification task, which has big significance for real-world applications. We are planning to extract features from the penultimate layer of the network and on the basis of their weighted chi-square distance perform face verification. Moreover, we would like to enrich training set by synthesizing new facial images in different poses using a 3D morphable model.

Acknowledgments. This work is supported by grant of the University of West Bohemia, project No. SGS-2016-039, by Ministry of Education, Youth and Sports of Czech Republic, project No. LO1506, by Russian Foundation for Basic Research, projects No. 15-07-04415 and 16-37-60100, and by the Government of Russian, grant No. 074-U01. Moreover, access to computing and storage facilities owned by parties and projects contributing to the National Grid Infrastructure MetaCentrum provided under the programme "Projects of Large Research, Development, and Innovations Infrastructures" (CESNET LM2015042), is greatly appreciated.

References

1. He, K., Zhang, X., Ren, S., Sun, J.: Deep residual learning for image recognition. In: Proceedings of 2016 IEEE Conference on Computer Vision and Pattern Recognition (CVPR), pp. 770–778 (2016)
2. Yi, D., Lei, Z., Liao, S., Li, Z.: Learning face representation from scratch. CoRR (2014)
3. Krizhevsky, A., Sutskever, I., Hinton, G.E.: ImageNet classification with deep convolutional neural networks. In: Proceedings of Advances in Neural Information Processing, pp. 1106–1114 (2012)
4. Taigman, Y., Yang, M., Ranzato, M., Wolf, L.: DeepFace: closing the gap to human-level performance in face verification. In: Proceedings of 2014 IEEE Conference on Computer Vision and Pattern Recognition, pp. 1701–1708 (2014)
5. Huang, G.B., Ramesh, M., Berg, T., Learned-Miller, E.: Labeled faces in the wild: a database for studying face recognition in unconstrained environments, pp. 07–49 (2007)
6. Sun, Y., Wang, X., Tang, X.: Deep learning face representation by joint identification-verification, pp. 1–9. CoRR (2014)
7. Sun, Y., Wang, X., Tang, X.: Deeply learned face representations are sparse, selective, and robust, pp. 2892–2900. CoRR (2014)
8. Lu, C., Tang, X.: Surpassing human-level face verification performance on LFW with Gaussian face. In: Proceedings of the Twenty-Ninth AAAI Conference on Artificial Intelligence, pp. 3811–3819 (2015)
9. Schroff, F., Kalenichenko, D., Philbin, J.: FaceNet: a unified embedding for face recognition and clustering. In: Proceedings of 2015 IEEE Conference on Computer Vision and Pattern Recognition (CVPR), pp. 815–823 (2015)
10. Klare, B.F., Klein, B., Taborsky, E., Blanton, A., Cheney, J., Allen, K., Grother, P., Mah, A., Burge, M., Jain, A.K.: Pushing the frontiers of unconstrained face detection and recognition: IARPA Janus Benchmark A. In: Proceedings of 2015 IEEE Conference on Computer Vision and Pattern Recognition (CVPR), pp. 1931–1939 (2015)
11. Masi, I., Rawls, S., Medioni, G., Natarajan, P.: Pose-aware face recognition in the wild. In: Proceedings of 2016 IEEE Conference on Computer Vision and Pattern Recognition (CVPR), pp. 4838–4846 (2016)
12. Russakovsky, O., Deng, J., Su, H., Krause, J., Satheesh, S., Ma, S., Huang, Z., Karpathy, A., Khosla, A., Bernstein, M., Berg, A.C., Fei-Fei, L.: ImageNet large scale visual recognition challenge. Int. J. Comput. Vis. (IJCV), pp. 211–252 (2015)
13. Kemelmacher-Shlizerman, I., Seitz, S.M., Miller, D., Brossard, E.: The MegaFace benchmark: 1 million faces for recognition at scale. In: Proceedings of the IEEE Conference on Computer Vision and Pattern Recognition (2016)
14. Jia, Y., Shelhamer, E., Donahue, J., Karayev, S., Long, J., Girshick, R., Guadarrama, S., Darrell, T.: Caffe: convolutional architecture for fast feature embedding (2014). arXiv preprint arXiv:1408.5093
15. Milborrow, S., Morkel, J., Nicolls, F.: The MUCT landmarked face database. In: Pattern Recognition Association of South Africa (2010)

Footstep Planner Algorithm for a Lower Limb Exoskeleton Climbing Stairs

Sergey Jatsun, Sergei Savin$^{(\boxtimes)}$, and Andrey Yatsun

Southwest State University, Kursk, Russia
savinswsu@mail.ru

Abstract. In this paper, a footstep planning algorithm for a lower limb exoskeleton climbing stairs is presented. The algorithm relies on having a height map of the environment, and uses two procedures: partial decomposition of the supporting surface into convex obstacle-free regions, and optimization of the foot step position implemented as a quadratic program. These two methods are discussed in detail in the paper, and the simulation results are shown. It is demonstrated that the algorithm works for different staircases, and even for the staircases with obstacles on them.

Keywords: Footstep planner algorithm · Lower limb exoskeletons · Quadratic programming · Climbing stairs

1 Introduction

Exoskeletons have been a focus of many studies in robotics for the past few decades because of the significant range of possible practical applications that this technology has. This variety of applications has already lead to a significant interest in the industry, resulting in a number of successfully working exoskeleton designs [1–3]. The lower limb exoskeletons are an especially important example of this development. They can serve to restore mobility to patients, and enable workers and soldiers to perform more physically demanding tasks [3].

At the same time, there is still a number of challenging problems for exoskeletons. One of the most important examples of such a problem is climbing stairs [4]. The challenges associated with climbing stairs include the problem of maintaining vertical balance of the mechanism, higher demands on the mechanical design and motors (as compared to walking on a horizontal plane) and the footstep planning problem [5, 6]. Here we consider the latter. A footstep planning algorithm should be able to produce a feasible sequence of steps using the information about the exoskeleton's position and the environment. The problem of processing the sensory data in order to construct a map of the environment, and find robot's location on it, is closely related to footstep planning [7]. This problem is usually being solved before the footstep planning algorithm can start working and is often associated with simultaneous localization and mapping methods or constructing the so-called height map [7, 8]. In this paper, we assume that the robot has access to the height map of the environment, and we focus on using this information in order to plan a sequence of steps.

© Springer International Publishing AG 2017
A. Ronzhin et al. (Eds.): ICR 2017, LNAI 10459, pp. 75–82, 2017.
DOI: 10.1007/978-3-319-66471-2_9

There are a few different approaches to footstep planning. In the paper [9] an optimization-based footstep planner is discussed. One of its features is the use of mixed-integer programming for planning and IRIS algorithm for the decomposing the supporting surface into convex regions (see [10] for details). Alternative approach based on stereographic projection is shown in [11]. In [12, 13] a footstep planner was realized using quadratic programming (QP) for different types of walking robots. In [14] a footstep planner for Honda's biped robot was designed based on the A* search algorithm. This planner does not guarantee optimality of the chosen path, but works at a satisfactory speed. We should note that significant progress in numerical techniques for solving quadratic programs has been made over the last decades, making the use of QP in real time control loops a viable solution. Examples of this can be found in works [15, 16]. In this paper, we use also use quadratic programming as a part of the footstep planning algorithm. Our algorithm also requires partial decomposition of the supporting surface (or its height map) into convex regions. The algorithm used for this decomposition is described in the following chapters.

2 Exoskeleton Description

In this paper, we focus on a footstep planner algorithm for a lower limb exoskeleton. We consider the exoskeleton ExoLite, which has two legs divided into four segments (hip, thigh, shin and foot). This exoskeleton has 10 actuated joints, one in each hip, thigh, and knee, and two in each foot [17–23]. Each joint is equipped with sensors for measuring joint angles and the exoskeleton's feet are additionally equipped with pressure sensors and inertial measurement units. A general view of the exoskeleton is shown in Fig. 1.

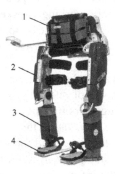

Fig. 1. General view of the ExoLite exoskeleton; 1 – torso link, 2 – thigh link, 3 – shin link, 4 – foot link

In papers [17–19] the task of controlling the robot while maintaining its vertical balance is discussed. Papers [20–23] are studying the problem of controlling the exoskeleton during the execution of prescribed trajectories, as well as the problem of tuning the controller.

The contact surface of the exoskeleton's feet has the form of a non-convex polygon. In this study, we consider an algorithm that finds a new placement position for the exoskeleton's feet. The new position should guarantee that the foot lies inside an obstacle-free region on the supporting surface. If we only consider obstacle-free regions that are convex, then we can modify this condition in the following way: the convex hull F_c of the points on the exoskeleton's foot should lie inside an obstacle-free region on the supporting surface. To show the equivalence of the two conditions we can observe that if all points on the exoskeleton's foot lie inside a convex region, then any convex combination of these points also lies inside that region, which in turn means that their convex hull lies inside the mentioned region.

In the following chapters, we assume that the robot has access to the height map of the supporting surface. The height map is a function $z_{map} = z_{map}(x, y)$ that returns a height z for a given point on the supporting surface with coordinates x and y.

3 Supporting Surface Partial Decomposition Algorithm

In this chapter, we consider an algorithm that we use to partially decompose the surface into convex obstacle-free regions. By obstacle-free region we mean a region where the heights of any two points are different by no more than ε, the height variation threshold. We say that the surface is partially decomposed to indicate that the procedure is meant to provide us with a few convex obstacle-free regions, but these regions do not need to tile the whole surface.

The first step of the algorithm is to generate a set of seed points $\Sigma = \{\sigma_i\}$, where $\sigma_i = [x_i \quad y_i]$ is a seed point with Cartesian coordinates x_i and y_i. To generate the set Σ we can use a grid, a set of random numbers or a low-discrepancy sequence of points, such as the Sobol sequence [24].

The second step is to take one seed point σ_i and construct an approximation of the convex obstacle-free region it belongs to. This is done in the following way. First we construct a sequence of n rays that originate from the point σ_i. The first ray can be chosen randomly, and every other ray in the sequence is constructed by rotating the previous one by $2\pi/n$ radians. We can parametrize these rays by a parameter ξ that represents the distance from the seed point σ_i. For the case when the first ray is chosen to lie along the positive x axis on the xy plane every point on each ray can be described by the next formula:

$$\chi_{i,j}(\xi) = \sigma_i + [\,\xi \cos(2j\pi/n) \quad \xi \sin(2j\pi/n)\,], \tag{1}$$

where $\chi_{i,j}$ is a point on the j-th ray and ξ is a variable that determines the distance from $\chi_{i,j}$ to the seed point σ_i. Then for each ray we find a point $v_{i,j}$ (where i is the index number of the seed point, and j is the index number of the ray, $j = \overline{1, n}$) closest to point σ_i out of all points on the ray which lie outside the obstacle-free region. His can be formulated as follows:

$$\begin{cases} v_{i,j} = \chi_{i,j}(\xi_{\min}) \\ \xi_{\min} = \min\{\xi : |z_{map}(\chi_{i,j}(\xi)) - z_{map}(\sigma_i)| > \varepsilon/2\} \end{cases}, \tag{2}$$

where $\xi_{\min}$ is the minimum value of ξ, such that the point $\chi_{i,j}$ lies on an obstacle. To determine whether or not $\chi_{i,j}(\xi)$ lies on an obstacle we use the height map z_{map} to check if the height difference between point $\chi_{i,j}$ and the seed point σ_i is greater than $\varepsilon/2$, which gives us a conservative estimation.

Thus we can construct an approximation of the convex obstacle-free region as a convex hull of the points $v_{i,j}$ (for a given i and all j). We denote this region as Ω_i:

$$\Omega_i = \text{Conv}\{v_{i,j}\} \quad j = \overline{1, n}. \tag{3}$$

We should note that the quality of this approximation depends the shape of the actual obstacle free region and on the number of rays n. If the region is convex, then as the number of rays grows the approximation will approach the exact shape of the region.

Because Ω_i is a convex hull of a finite number of points it is a polygon with vertices $v_{i,j}$, and as any convex polygon it can be represented as a set of linear inequalities:

$$\Omega_i = \{\mathbf{r} : \mathbf{A}_i\mathbf{r} \le \mathbf{b}_i\}, \tag{4}$$

where $\mathbf{A}_i$ and $\mathbf{b}_i$ are a matrix and a vector that correspond to a linear inequality representation of Ω_i, and $\mathbf{r}$ is a radius vector for a point in the polygon Ω_i. For the discussion of algorithms for computing $\mathbf{A}_i$ and $\mathbf{b}_i$ from the given set of vertices $v_{i,j}$, see [25].

The third step of the algorithm is the improvement of the obstacle-free region approximation. Since Ω_i was chosen as a convex hull of the points $v_{i,j}$, it can contain some of these points in its interior. During this step we add new inequalities to the system $\mathbf{A}_i\mathbf{r} \le \mathbf{b}_i$ and construct a new convex polygon Ω_i^*, such that every point $v_{i,j}$ is either outside Ω_i^* or lie on its boundary.

Let us assume that a vertex defined by its radius vector $\mathbf{v}_{in}$ lies in the interior of Ω_i. Then, we can add the following linear inequality to the system $\mathbf{A}_i\mathbf{r} \le \mathbf{b}_i$ to make it lie on the boundary of a new polygon Ω_i^*:

$$\begin{cases} \mathbf{a}_{new}^T\mathbf{r} \le b_{new} \\ \mathbf{a}_{new} = (\mathbf{v}_{in} - \boldsymbol{\sigma}_i)/\|\mathbf{v}_{in} - \boldsymbol{\sigma}_i\|, \\ b_{new} = \|\mathbf{v}_{in}\| + \mathbf{a}_{new}^T\boldsymbol{\sigma}_i \end{cases} \tag{5}$$

where $\boldsymbol{\sigma}_i$ is the radius vector defining the position of the point σ_i. Then the new polygon Ω_i^* is defined as follows:

$$\Omega_i^* = \left\{\mathbf{r} : \begin{bmatrix} \mathbf{A}_i \\ \mathbf{a}_{new}^T \end{bmatrix} \mathbf{r} \le \begin{bmatrix} \mathbf{b}_i \\ b_{new} \end{bmatrix}\right\}. \tag{6}$$

We iteratively add new constraints to the polygon Ω_i^* until all vertices $v_{i,j}$ are outside Ω_i^* or lie on its boundary.

4 Quadratic Programming-Based Footstep Planning Algorithm

In this chapter, we discuss a footstep planning algorithm for climbing stairs based on quadratic programming. We denote as $\mathbf{r}_d = [x_d \quad y_d]^T$ a desired step. A desired step is determined by the user, and the planning algorithm tries to find a feasible step that would be as close as possible to $\mathbf{r}_d$. We formulate it as the following optimization problem:

$$\begin{aligned} &\text{minimize} \quad \mathbf{r}_s^T \mathbf{r}_s \\ &\text{subject to} \quad \mathbf{A}_i^* (\mathbf{r}_d + \mathbf{r}_s + \mathbf{r}_{ej}) \leq \mathbf{b}_i^* \end{aligned} \qquad (7)$$

where $\mathbf{r}_s = [x_s \quad y_s]^T$ is a shift from the desired foot position $\mathbf{r}_d$, $\mathbf{r}_{ej}$ is the relative position of the j-th vertex of the polygon F_c, which corresponds to a vertex of the exoskeleton's foot, and $\mathbf{A}_i^*$ and $\mathbf{b}_i^*$ are a matrix and a vector in the linear inequality representation of the obstacle-free region Ω_i^*. This problem is solved for all obstacle-free regions Ω_i^* that were found using the previously discussed decomposition procedure, and then the solution with the smallest value of the cost function is chosen.

This algorithm can be modified by changing the cost function to be a quadratic form of $\mathbf{r}_s$. Then by using an appropriate quadratic form matrix we can chose the preferred direction of the shift $\mathbf{r}_s$.

5 Simulation Results

In this chapter, we look at simulation results obtained using the presented algorithm. First, we consider two different staircases with pitch angles $\pi/6$ and $\pi/12$. Both staircases have a rise height of 0.12 m. The desired step length was chosen to be 0.32 m. The obtained footstep plans are shown in Fig. 2.

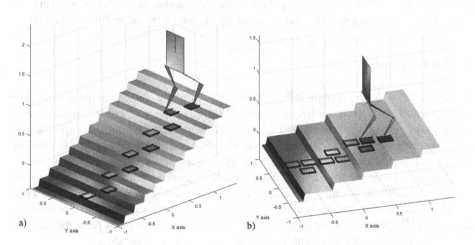

Fig. 2. The exoskeleton climbing staircases with pitch angles: (a) $\pi/6$, (b) x/12

We can note that for the steeper staircase (Fig. 2a) the footstep plan has hit every step of the staircase exactly once. For the other staircase, the footstep plan has hit every step three times. This happened because of the greater length of the steps. By doubling the desired step length, we can obtain a footstep plan that would skip every other step.

The proposed algorithm can also handle cases when there are additional obstacles on the staircase. Figure 3 shows the footstep plan generated for the case when the staircase has a pitch angle of $\pi/8$, and there are 3 obstacles lying on the stairs.

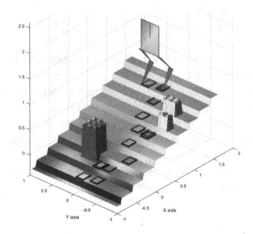

Fig. 3. The exoskeleton climbing a staircase with obstacles

We should note that the third step of the partial decomposition algorithm is mainly used to handle this type of problem. Stairs with no obstacles naturally provide convex regions, so only the first two steps of the decomposition algorithm are needed. On the other hand, addition of the third step to the decomposition algorithm allows to handle the problem of feet collision, by checking if the vertices of the stationary foot lie in the interior of the obstacle-free regions Ω_i^*, and augmenting Ω_i^* with new inequalities if they do.

6   Conclusions

In this paper, a footstep planning algorithm for climbing stairs was presented. The algorithm relies on having a height map of the supporting surface. It can be decomposed into two independent methods: first is the method for partial decomposition of the supporting surface into convex regions and the second is an algorithm for finding an optimal foot placement in these regions using quadratic programming.

It is shown that the footstep planning algorithm works for different staircases, and that it can handle the situation when there are obstacles lying on the stairs. It shows that the algorithm can be used as an alternative to the method presented in [26] for obstacle avoidance. The version of the algorithm presented in the paper plans for only one step ahead, but it is possible to extend it for planning multiple steps ahead.

Acknowledgements. Supported from grant of the President of the Russian Federation for young scientists MK-2701.2017.8.

References

1. Barbareschi, G., Richards, R., Thornton, M., Carlson, T. and Holloway, C.: Statically vs dynamically balanced gait: analysis of a robotic exoskeleton compared with a human. In: 2015 37th Annual International Conference of the IEEE Engineering in Medicine and Biology Society (EMBC), pp. 6728–6731 (2015). doi:10.1109/EMBC.2015.73199372015
2. Kazerooni, H., Racine, J.L., Huang, L., Steger, R.: On the control of the berkeley lower extremity exoskeleton (BLEEX). In: Proceedings of the 2005 IEEE International Conference on Robotics and Automation, ICRA 2005, pp. 4353–4360 (2005). doi:10.1109/ROBOT. 2005.1570790
3. Ferrati, F., Bortoletto, R., Menegatti, E., Pagello, E.: Socio-economic impact of medical lower-limb exoskeletons. In: 2013 IEEE Workshop on Advanced Robotics and its Social Impacts (ARSO), pp. 19–26 (2013). doi:10.1109/ARSO.2013.6705500
4. Fu, C., Chen, K.: Gait synthesis and sensory control of stair climbing for a humanoid robot. IEEE Trans. Ind. Electron. **55**(5), 2111–2120 (2008). doi:10.1109/TIE.2008.9212052008
5. Lum, H.K., Zribi, M., Soh, Y.C.: Planning and control of a biped robot. Int. J. Eng. Sci. **37** (10), 1319–1349 (1999). doi:10.1016/S0020-7225(98)00118-9
6. Figliolini, G., Ceccarelli, M.: Climbing stairs with EP-WAR2 biped robot. In: IEEE International Conference on Robotics and Automation, Proceedings 2001 ICRA, vol. 4, pp. 4116–4121 (2001). doi:10.1109/ROBOT.2001.933261
7. Fallon, M., Kuindersma, S., Karumanchi, S., Antone, M., Schneider, T., Dai, H., D'Arpino, C.P., Deits, R., DiCicco, M., Fourie, D., Koolen, T.: An architecture for online affordance-based perception and whole-body planning. J. Field Robot. **32**(2), 229–254 (2015). doi:10.1002/rob.21546
8. Thrun, S., Leonard, J.J.: Simultaneous localization and mapping. In: Siciliano, B., Khatib, O. (eds.) Springer Handbook of Robotics, pp. 871–889. Springer, Heidelberg (2008). doi:10. 1007/978-3-540-75388-9_3
9. Kuindersma, S., Deits, R., Fallon, M., Valenzuela, A., Dai, H., Permenter, F., Koolen, T., Marion, P., Tedrake, R.: Optimization-based locomotion planning, estimation, and control design for the Atlas humanoid robot. Auton. Robots **40**(3), 429–455 (2016). doi:10.1007/ s10514-015-9479-3
10. Deits, R., Tedrake, R.: Computing large convex regions of obstacle-free space through semidefinite programming. In: Akin, H.L., Amato, N.M., Isler, V., Stappen, A.F. (eds.) Algorithmic Foundations of Robotics XI. STAR, vol. 107, pp. 109–124. Springer, Cham (2015). doi:10.1007/978-3-319-16595-0_7
11. Savin, S.: An algorithm for generating convex obstacle-free regions based on stereographic projection. In: Proceedings of the 13th Siberian Conference SIBCON-2017 (2017, in publishing)
12. Feng, S., Whitman, E., Xinjilefu, X., Atkeson, C.G.: Optimization-based full body control for the DARPA robotics challenge. J. Field Robot. **32**(2), 293–312 (2015). doi:10.1002/rob. 21559
13. Savin, S., Vorochaeva, L.Y.: Footstep planning for a six-legged in-pipe robot moving in spatially curved pipes. In: proceedings of The 13th Siberian Conference SIBCON-2017 (2017, in publishing)

14. Chestnutt, J., Lau, M., Cheung, G., Kuffner, J., Hodgins, J., Kanade, T.: Footstep planning for the Honda ASIMO humanoid. In: Proceedings of the 2005 IEEE International Conference on Robotics and Automation, ICRA 2005, pp. 629–634 (2005). doi:10.1109/ROBOT.2005.1570188

15. Kuindersma, S., Permenter, F., Tedrake, R.: An efficiently solvable quadratic program for stabilizing dynamic locomotion. In: 2014 IEEE International Conference on Robotics and Automation (ICRA), pp. 2589–2594 (2014). doi:10.1109/ICRA.2014.6907230

16. Wang, Y., Boyd, S.: Fast model predictive control using online optimization. IEEE Trans. Control Syst. Technol. **18**(2), 267–278 (2014). doi:10.1109/TCST.2009.20179342010

17. Panovko, G.Y., Savin, S.I., Yatsun, S.F., Yatsun, A.S.: Simulation of exoskeleton sit-to-stand movement. J. Mach. Manuf. Reliab. **45**(3), 206–210 (2016). doi:10.3103/S1052618816030110

18. Jatsun, S., Savin, S., Yatsun, A., Gaponov, I.: Study on a two-staged control of a lower-limb exoskeleton performing standing-up motion from a chair. In: Kim, J.-H., Karray, F., Jo, J., Sincak, P., Myung, H. (eds.) Robot Intelligence Technology and Applications 4. AISC, vol. 447, pp. 113–122. Springer, Cham (2017). doi:10.1007/978-3-319-31293-4_10

19. Jatsun, S.F.: The modelling of the standing-up process of the anthropomorphic mechanism. In: Assistive Robotics: Proceedings of the 18th International Conference on CLAWAR 2015, pp. 175–182 (2015). doi:10.1142/9789814725248_0024

20. Jatsun, S., Savin, S., Lushnikov, B., Yatsun, A.: Algorithm for motion control of an exoskeleton during verticalization. In: ITM Web of Conferences, vol. 6, pp. 1–6 (2016). doi:10.1051/itmconf/20160601001

21. Jatsun, S., Savin, S., Yatsun, A.: Parameter optimization for exoskeleton control system using Sobol sequences. In: Parenti-Castelli, V., Schiehlen, W. (eds.) ROMANSY 21 - Robot Design, Dynamics and Control. CICMS, vol. 569, pp. 361–368. Springer, Cham (2016). doi:10.1007/978-3-319-33714-2_40

22. Jatsun, S., Savin, S., Yatsun, A.: Comparative analysis of iterative LQR and adaptive PD controllers for a lower limb exoskeleton. In: 2016 IEEE International Conference on Cyber Technology in Automation, Control, and Intelligent Systems (CYBER), pp. 239–244 (2016). doi:10.1109/CYBER.2016.7574829

23. Jatsun, S., Savin, S., Yatsun, A., Postolnyi, A.: Control system parameter optimization for lower limb exoskeleton with integrated elastic elements. In: Advances in Cooperative Robotics, pp. 797–805 (2017). doi:10.1142/9789813149137_0093

24. Sobol, I.M.: On the distribution of points in a cube and the approximate evaluation of integrals. Zhurnal Vychislitel'noi Matematiki i Matematicheskoi Fiziki **7**(4), 784–802 (1967)

25. Bremner, D., Fukuda, K., Marzetta, A.: Primal-dual methods for vertex and facet enumeration. Discret. Comput. Geom. **20**(3), 333–357 (1998). doi:10.1007/PL000093891998

26. Jatsun, S., Savin, S., Yatsun, A.: Study of controlled motion of an exoskeleton performing obstacle avoidance during a single support walking phase. In: 20th International Conference on System Theory, Control and Computing (ICSTCC), pp. 113–118 (2016). doi:10.1109/ICSTCC.2016.7790650

Synthesis of the Behavior Plan for Group of Robots with Sign Based World Model

Gleb A. Kiselev[1] and Aleksandr I. Panov[1,2(✉)]

[1] Federal Research Center "Computer Science and Control",
Russian Academy of Sciences, Moscow, Russia
{kiselev,pan}@isa.ru
[2] National Research University Higher School of Economics, Moscow, Russia

Abstract. The paper considers the task of the group's collective plan intellectual agents. Robotic systems are considered as agents, possessing a manipulator and acting with objects in a determined external environment. The MultiMAP planning algorithm proposed in the article is hierarchical. It is iterative and based on the original sign representation of knowledge about objects and processes, agents knowledge about themselfs and about other members of the group. For distribution actions between agents in general plan signs "*I*" and "*Other*" ("*They*") are used. In conclusion, the results of experiments in the model problem "Blocksworld" for a group of several agents are presented.

Keywords: Multiagent planning · Sign · Behavior planning · Task planning · Robots · Group of robots · Sign based world model

1 Introduction

Collective robotics mind currently attract attention of specialists in artificial intelligence (AI), as well as experts in the actual robotics. Great success has been achieved in team games (i.e. RoboCup [10]) and joint movement of unmanned vehicles (swarms and UAV groups) [7,8,19]. However, the task of allocating roles in a group of autonomous agents that solve more than one of the specific problem, is universal, i.e. able to learn in the new stating the statements and simultaneously taking into account the opportunities and training of others agents that are integral parts of the solution was much more complicated and good-enough solution still does not appear. The main efforts to solve this problem of robotic system that would allow it not only to function sport, but also to learn in it, to build conceptual plans of behavior and exchange messages with other members of the coalition [2,6,18]. One of the key the subtask in creating such cognitive resources is the selection and use of method of knowledge representation and the basis for the synthesis behavior plan.

In this paper we present the original MultiMAP algorithm - behavior plan synthesis of a group of intelligent agents that represent robotic systems with different manipulators, which constrain their ability to interact with objects of

© Springer International Publishing AG 2017
A. Ronzhin et al. (Eds.): ICR 2017, LNAI 10459, pp. 83–94, 2017.
DOI: 10.1007/978-3-319-66471-2_10

the external environment. For example, some agents can operate only with big objects and others only with small ones. Such a difference in functionality leads to the fact that some tasks can not be solved only by one agent and encourage him to consider the possible actions of other agents. To represent knowledge about processes and objects of the external environment, knowledge about himself and other agents in real work uses a model of sign based world model [13–15], which relies on the psychological theory of human activity [9,11]. The base element of sign world model is a four-component structure - a sign that combines both declarative and procedural knowledge about the object, process or subject activities. The agent keeps the idea of himself and his abilities in the sign "I", and information about the abilities of other agents - in the corresponding signs "A_1", "A_2", etc.

The proposed planning algorithm is iterative and hierarchical. Convergence of the iterations of planning from the declared event case, if the knowledge or skills of the agent is enough to build a complete plan. Hierarchy is that conceptual actions the agent at the beginning tries to use, specifying them if necessary. The model of sign based world model allows an agent to include in the plan both his own actions and the actions of other agents. Also, the agent remembers all his successful attempts to achieve goals and subgoals, using the accumulated experience in new situations, which allows planning on the precedents, the decreasing time, the execution of the plan and the length of the resulting chains of actions.

The paper is organized as follows. In Sect. 2 complete the statement of the planning in a group of intelligent agents problem and a short description of the presentation tasks for agents (ML-PDDL and PDDL3). In Sect. 3 a brief overview of the planning in multi-agent systems and comparison with the present work. In Sect. 4 presented implementation of algorithm in the group of agents possessing the sign based world model, and a description of the model problem "Blocksworld" is given. In Sect. 5 the results of model experiments and their discussion are given.

2   Problem Statement

In this work, an intelligent agent will be understood as hardware or a software system that has the autonomy, reactivity, activity, and commutativity properties. These properties allow the agent to interact with the environment that includes various types of objects. The plan P of the agent A_k is called the sequence of actions $P(A_k) = \langle a_1(A_{i1}), a_2(A_{i2}), \ldots \rangle$, obtained as a result of planning algorithm work, where A_{ij} is the agent performing the action a_j. The plan is formed by agent based on the goal, information about the current state of the environment and the dynamics of its change. In the present paper we consider the simplest case where agents modelling plans in tasks which have only the one possible choice of the roles distribution. This allows us not to consider the problems of coordinating plans.

We describe the task and the planning domain using languages PDDL3 and ML-PDDL [3], which uses the predicate calculus of the first order. The planning task $T = \langle AO, S, F, C \rangle$ consists of a set of agents and objects of the environment OE (example in Table 1 - field: objects), initial S and final F states of the environment (fields *:init* and *:goal*) and the constraints on the actions of agents C (field *:constraints*). The initial state S consists of a predicates set which are true in the initial situation, before the action is initiated by the agent, F includes the predicates that are true in the goal situation of the problem. The set C describes the constraints on the applicability of an action in the context of agent properties. For example, agent A_1 only works with small objects, and A_2 only with big ones.

The domain of the problem $D = \langle V, TO, A \rangle$ includes descriptions of predicates V, types of TO objects, and actions of agents A. The type of objects defines a class-subclass relation in the agent's knowledge base. Predicates describe some statement about an object (for example, a predicate $\langle blocksize \rangle$: a big block). The planning domain includes non-specific predicates that describe the type of relationship between an object of a certain type and its specific property. The planning task includes the specified predicates, where an abstract type of substituted coefficients is found. The description of the action $a = \langle n, Cond, Eff \rangle$ includes the name of the action n, a list of its $Cond$ preconditions and Eff effects. The list of preconditions includes predicates that form the condition for applying this action, and the effect listings consists of predicates, which meanings became true after applying the action. The effect also includes a set of predicates with the NOT key, which denote those predicates whose meaning has ceased to be true.

In the article the task is transformed from the description presented in PDDL to a sign based world model. The loading the world model process of an agent is explicitly described in Sect. 4.2.

3 Related Works

Planning in a group of agents is a sufficiently developed direction in the theory of multi-agent systems. We note a number of papers using a close formulation of the problem.

In the article [5] the approach directed on improvement of productivity of agents in successive manipulation problems is considered. A disconnected with the agent optimization-based approach to problem solving planning and motion adjustment is described, including elements of symbolic and geometric approaches. The algorithm generates auxiliary actions of the agent, which allows other participants to reduce cognitive and kinematic load during the task execution being in the coalition of agents. The task domain is described using symbolic predicates with the addition of special functions to implement geometric constraints in the algorithm. The target state is described by abstract predicates, which make it possible to indistinctly represent individual components. Tasks of this type can be solved using the MultiMAP scheduling algorithm described in this paper, personalizing the process constructing a plan by using knowledge about the current preferences of each of the team member and made by each one.

Table 1. Example of domain description and planning task

Planning domain	Planning task
(define (domain domain-name)	(define (problem BLOCKS-4-0)
(:requirements :typing :multi-agent)	(:domain blocks)
(:types	(:objects
agent object1 - object	a - object1
)	b - object1
(:predicates	ag1 agent
(predicate-1 ?ob1 object1 ?ag agent)	ag2 - agent)
(predicate-2 ?ob1 object1)	(:init
(predicate-3 ?ag agent))	(predicate-1 a ag1)
(:action action-name	(predicate-1 b ag2))
:parameters (?ob1 object1 ?ag agent)	(:goal
:precondition (and	(and
predicate-1 ?ob1 ?ag))	(predicate-2 a)
:effect (and	(predicate-3 ag1)
(not (predicate-1 ?ob1 ?ag))	(predicate-2 b)
(predicate-2 ?ob1)	(predicate-3 ag2))))
(predicate-3 ?ag agent)))	(:constraints
	(and
	(and (always (forall (?ob1 - object1)
	(implies (predicate-2 ?a) (predicate-1 ?
	x ?ag1)))))

The article [12] considers a multi-agent approach to the use of the LAMA planning algorithm [20]. The planning algorithm described in the article builds plans of various levels, where the plan of the highest level consists only of public actions of agents. Each public action is interpreted by the agent as a private action plan based on facts internal to the agent. A high-level plan is considered to be valid if each agent draws up an action plan independently from other agents until the private preconditions for public action are reached, which he performs in a high-level plan. The means of representing knowledge in the article is the logic of the predicate calculus of the first order.

The article [1] describes the algorithms of weak and strong privacy, created to realize the possibility of hiding personal actions that make up the process of performing a public action within the framework of a multi-agent approach to the construction of team of agents action plan. To describe the domain and the planning task, a multi-agent version of the PDDL language is used, which allows manipulating with private actions and facts. In this paper, each of the agents plans make actions within their own world map processes, which implies the use of private, personal agent actions. However, the notions of limitations of possible actions of other agents are part of the picture of the world of the planning agent, which he created on the basis of available data on other agents.

4   Method and MultiMAP Algorithm

4.1   Sign Based World Model

In this paper the model of the sign based world model is used as the main way of knowledge representation, the basic element of which is a four-component structure, called the sign [14,16]. The sign can represent both a static object and an action. The sign is given by the name and contains the components of the image, significance, and personal meaning. The image component contains the characteristic features of the represented object or process. The significance component represents generalized usage scenarios for the collective of agents. The component of the personal meaning of the sign determines the role of the object in the actions of the subject committed by him with this object. Personal meanings of the sign are formed in the process of the subject's activity and are the concretization of scenarios from the meaning of this sign. Personal meanings reveal the preferences of the subject of activity, reflect the motive and emotional coloring of the actions.

The components of the sign consist of special semantic (causal) networks at the nodes of which the so-called causal matrices are located. Each causal matrix represents a sequence of lists (columns) of attributes of a given sign component. The attributes are either elementary data from sensors or references to the corresponding signs. For example, the causal matrix of the sign *on* consists of two columns: the left column contains the sign of the *blockX*, and the right sign the *blockY*, which indicates that the *blockX* is *on* the *blockY*. Using each of the three resulting causal networks, it is possible to describe a series of semantic relations on a set of signs. So among the relations on a set of signs on a network of significance there is a class–subclass relation, when for one sign designating some role there can be several signs playing this role, forming its causal matrix on a network of significance.

Formally, the s symbol is a tuple of four components: $\langle n, p, m, a \rangle$, where is the name of the sign, p is the image of the sign corresponding to the node $w_p(s)$ of the causal network on the images, m is the sign significance corresponding to the node $w_m(s)$ of the causal network of significances, a - the personal meaning of the sign, corresponding to the node $w_a(s)$ of the causal network on the senses. R_n - relations on the set of signs, and θ - operations on the set of signs, obtained on the basis of fragments of causal networks, to which the corresponding sign components belong. A tuple of five elements $\langle W_p(s), W_m(s), W_a(s), R_n, \theta \rangle$ is a model of a semiotic network.

The model of the sign based world model is used in this paper as the basic way of representing knowledge for building collective plans. As part of the process of plan finding in a sign based world model, the reverse planning process (from the target situation) is carried out, described in detail in Sect. 4.2. Agents perform various actions based on their personal meanings and try to reach the initial situation. Knowledge of the capabilities of the planning agent and other agents is represented by nodes on the causal networks of the personal meanings of the sign "I" and the signs "A_1", "A_2" etc. for each agent, respectively, which allows agents to plan their actions based on the criteria for the possibility of applying actions by the planning agent, or from the possibility of applying the actions of someone from other agents of the group. Like any sign, the causal matrix of sign "I" consists of the personal meanings of the agent, carrying out the planning process, its image and significances. The image of the agent and other members of the group is his main characteristic, important for the recognition of other agents by sensor data, so we have omitted this component. The significance of the "I" sign and the signs of other agents are generalized scenarios (actions) in which an agent can act as an entity either directly or through its classes. All actions that agent A_i can perform are represented in his personal meanings and are partially specified value whose action roles are prefilled in accordance with the block of constraints $C(A_i)$ presented in the planning task T.

The signs of other agents are related by a class-subclass relationship to the abstract sign "$They$". Agent signs include agent views of the remaining agents, derived from a general description of the planning task.

4.2 MultiMAP Algorithm

The planning process in the sign based world model is realized with MAP-algorithm [17,19] and goes in the opposite direction: from the final situation to the initial one. The input of the algorithm is a description of the task $T_{agent} = \langle N_T, S, Sit_{start}, Sit_{goal} \rangle$ where N_T - the task identifier, S - the set of signs, including the sign of the current agent S_I, and the signs of other agents $S_{Ag1}, S_{Ag2}, S_{Ag3}, \ldots$. $Sit_{start} = \langle \emptyset, \emptyset, a_{start} \rangle$ - the initial situation is modeled by a sign with meaning $a_{start} = \{z^a{}_{start}\}$, $Sit_{goal} = \langle \emptyset, \emptyset, a_{goal} \rangle$ - the target situation with meaning $a_{goal} = \{z^a{}_{goal}\}$.

Input: description of the planning domain D, description of the planning task T
maximum depth of iterations i_{max}
Output: $plan$

1. For agent in agents:
2. $T_{agent} := GROUND(P, A)$
3. $Plan := MAPSEARCH(T_{agent})$
4. function $MAPSEARCH(T_{agent})$
5. $z_{cur} := z^a_{goal}$
6. $z_{start} := z^a_{start}$
7. $Plans := MAPITERATION(z_{cur}, z_{start}, \emptyset, 0)$
8. $\{Plan_0, Plan_1, \ldots\} = SORT(PLANS)$
9.   return $Plan_0$
10. function $MAPITERATION(z_{cur}, z_{start}, Plan_{cur}, i)$
11.   if $i > i_{max}$ then:
12.   return $\emptyset$
13. $Act_{chains} = getsitsigns(z_{cur})$
14.   for chain in Act_{chains}:

15. $Act_{signif} = getactions(chain)$
16.   for pm_{signif} in Act_{signif}:
17. $Ch = openaction(pm_{signif})$
18. $meanings = generatemeanings(Ch, agents)$
19. $checked = activity(z_{cur}, meanings)$
20. $candidates = metaactivity(checked)$
21.   for candidate, agent in candidates:
22. $z_{cur+1} = timeshiftbackwords(z_{cur},$
23. $candidate)$
24. $plan.append(candidate, agent)$
25.   if $z_{cur+1} \in z_{start}$:
26.   return $plan$
27.   else: $Plans := MAPITERATION(z_{cur+1},$
28. $z_{start}, plan, i + 1)$

The algorithm begins with the process of symbol grounding (loading the sign based world model of the agent) GROUND, in which corresponding signs and relations are created for all objects, roles, predicates and actions in them in the causal networks of significances and personal meanings. For the initial and final situations, the signs Sit_{start} and Sit_{goal} are created and nodes of causal networks of personal meanings $w_a(Sit_{start})$ and $w_a(Sit_{goal})$ are defined. In general, the T_s task is the result of the symbol grounding procedure - the formation of world model from the original descriptions of the planning domain D and the planning task T.

The result of the implementation of the MultiMAP algorithm is a plan $Plan = \{\langle z_{s1}^a, z_{p1}^a, S_x \rangle, \langle z_{s2}^a, z_{p2}^a, S_x \rangle, \ldots, \langle z_{sn}^a, z_{pn}^a, S_x \rangle\}$ - sequence of the length of n pairs $\langle z_{si}^a, z_{pi}^a, S_x \rangle$, where z_{si}^a is the causal matrix of a certain network node on the personal meanings representing the i planning situation, z_{pi}^a - the causal matrix of some personal meaning representing the action applied in the z_{si}^a situation, and S_x the agent sign, $x \in \{I, Ag_1, Ag_2, Ag_3, \ldots\}$ depending on the possibility actions by the agent himself or by other agents. In this case, the situation z_{si+1}^a is the result of the action z_{pi}^a, in the sense that is revealed later in the discussion of the algorithm, $z_{s1}^a := z_{start}^a$ - the causal matrix, corresponding to the meaning of the initial situation, $z_{sn}^a := z_{goal}^a$ - the causal matrix corresponding to the meaning of the target situation.

For each agent, the planning algorithm starts with the process of signification. In this process, signs are generated from the data received by the agent through domain analysis and planning tasks. The process of signification is the process of creating a of the subject's world model, within the framework of this process, the networks of significances and personal meanings of the agent are formed from causal matrixes of signs, signs "I" and "$They$" are formed.

The process of signification begins with the creation of signs of agents and objects and with the creation of corresponding causal matrices on a network of significances. After this, signs and causal matrices of object types are created, procedural signs for predicates and actions. There is a formation of class-subclass relations between different object signs. In the causal matrix on the network of personal meanings of the sign "I" includes the sign of the agent performing the planning process, the signs of other agents are included in the causal matrix of the sign "They". Procedural and object signs that satisfy the constraints of agents are included in causal matrices on the network of personal meanings of the agent's sign. In the causal matrix on the network of personal meanings of the sign "I" are included constraints related to the activities of the planning agent, with respect to the available for use object and procedural signs. Causal matrices are created on the network of personal meanings of the signs of the initial and final situation, the corresponding object signs are included in them.

Steps 4–26 describe an algorithm for constructing an action plan for each of the agents. After this, the $MAPITERATION$ function described in clauses 10–26 is executed. At the first step of the recursively called $MAPITERATION$ function, the current recursion step is compared with the maximum iteration step i_{max} - if it is exceeded, the function returns an empty set. In step 13, all the characters appearing in the description of the current situation ($getsitsigns$) are received. In step 15 - the generation of all procedural marks ($getactions$), in step 17, a list of all object signs included in the causal matrices of procedural signs ($openaction$) is created. In steps 18–20, new nodes of the causal network of the personal senses of procedural signs are searched for, into the causal matrices of which the signs of the target situation enter. From point 21 to point 26, the process of creating a new situation occurs on the basis of the signs entering into the causal matrices of the received procedural signs and signs entering into the causal matrix of the current situation. In step 26, a recursive call to the

MAPITERATION function occurs if the new causal matrix created in step 22 is not included in the initial situation matrix, otherwise, the function returns the found plans.

4.3   Model Example

An example of work of the presented algorithm can be the solution of planning task in the well-known formulation of the "Blocksworld" [4] task. Problem domain is described in the multi-agent version of the PDDL language and serves as the initial data for the formation of the sign based world model. World model consist of object signs such as the sign of the *"block"* and procedural signs *"on"*, *"holding"*, *"pick-up"*, *"unstack"*, etc. The agent is specified by a special object with type agent. Blocks have different sizes (predicate *"size"*), and agents are available action *"wait"*, which contain empty sets of conditions and effects.

Here is an example of a MultiMAP planning algorithm solving a problem in which, in the initial situation, the four blocks *"a"*, *"b"*, *"c"* and *"d"* are on the table, the signs of each of these block includes in the causal matrix of the *"ontable"* sign, the block *"a"* and *"c"* are big, that is, the sign *"size"* is replaced by the sign *"big"* when forming the causal matrix of the procedural sign *"blocktype"* using the signs of these blocks in the process of signification, while the other two blocks are small and the sign *"size"* is replaced by the sign *"small"*. Two agents that operate with different types of blocks have empty manipulators and blocks are clear, that is, the signs of the blocks are included the causal matrices of the personal meaning of the sign *"clear"*. The result of solving this planning task are plans of each of the agents for the construction of the tower *"a-b-c-d"*, in which the sign of the block *"d"* is included the causal matrix of the procedural sign *"ontable"*, the causal matrix of the sign *"clear"* includes the sign of the block *"a"*, and the remaining signs form causal matrices of the procedural sign *"on"*. For each agent in the causal matrix, on a network of personal meanings includes procedural and object signs that satisfy the constraints of the problem.

In Fig. 1 shows a fragment of the causal matrix on the significance network of the procedural sign *"stack"*. A similar network displays a sequence of class-subclass relations and a role-sign relations for each of the action objects in the MAP-planning algorithm. Thus, the sign *"block?x"* during the formation of the causal matrix of procedural signs is replaced by the sign of any of the blocks, which does not contradict the constraints of the problem described in the set C.

At the first stage of the main iteration of the algorithm, set of precedents are obtained, empty in the simplest case. At the next stage, the process of obtaining object signs included in the current situation and defining the procedural signs *"stack"*, *"unstack"*, *"put-down"*, *"pick-up"* and *"wait"*.

Next, the formation of causal matrices of procedural signs occurs by replacing the signs of roles with object signs, in Fig. 1 shows the process of constructing the matrix of the procedural sign *"unstack"* from the object sign *"a"*. At this stage, the applicability of the resulting causal matrices of the procedural signs is checked by screening out those causal matrices that do not fit the constraints of the problem, or do not satisfy the selection heuristic. Constraints of the task are related to the possibility of applying the actions of the agent, which are described by the presence of appropriate procedural and object signs in the causal matrix on the network of personal meanings of the agent's sign. In Fig. 1, the causal matrices that are involved in the formation of the causal matrix on the network of personal meanings of the procedural sign "*unstack*" are highlighted in green, under the condition that the signs "*a*" and "*b* are included in the required causal matrix. Since the block "*a*" is big by the condition of the problem (the object sign "*a*" included in the causal matrix of the significances of the procedural sign "*blocktype*" along with the object sign "*big*"), taking into account the constraints of the problem, only the agent "*A*$_1$" can perform manipulations with it.

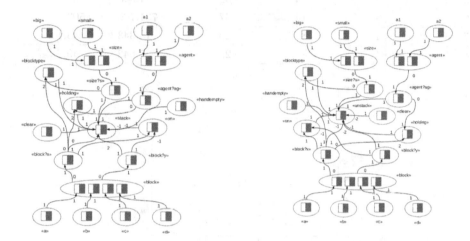

Causal matrix of the procedural sign   The process of obtaining the sign of the
"*stack*" on significance network   procedural matrix of the "*unstack*" action

Fig. 1. Network representations of actions "*stack*" and "*unstack*"

At the final stage of the iteration, a new causal matrix z_{next} is created on the network of personal meanings, representing the following planning situation. As a result of the algorithm, plans are created by each of the agents:

$A_1 : (pick - upc, I), (stackcd, I), (pick - upb, A_2), (stackbc, A_2), (pick - upa, I), (stackab, I).$

$A_2 : (pick - upc, A_1), (stackcd, A_1), (pick - upb, I), (stackbc, I), (pick - upa, A_1), (stackab, A_1).$

5   Experiments

Within the model problem presented in Sect. 4.3, experiments were performed describing the problems solved by the MultiMAP planning algorithm. The results of experiments are described in Table 2.

Based on the obtained data, the graphs of memory usage and time spent were constructed depending on the complexity of the tasks (Fig. 2).

Table 2. Results of the conducted experiments with the MAP-algorithm in the context of solving the problems of the "Blocksworld" task (WM - sign based world model)

#	Amount of agents	Amount of blocks (task complexity)	Plans length	Amount of signs in agents WM	Plans time counting	Amount of memory for each plan
1	2	4	6	29	9.189	34.8
2	2	5	8	30	20.153	46.5
3	3	5	8	32	61.627	83.4
4	3	6	10	33	114.927	130.6
5	3	8	14	35	321.948	307.6
6	4	8	14	37	731.760	503.1
7	3	4	8	29	15.343	143.1
8	3	4	8	29	7.842	49.7

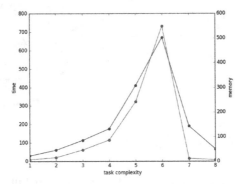

Fig. 2. Graphs of the dependence of the memory used and the timing of the plans processing relative to the complexity of the problem. "a"

According to the data shown in Fig. 2, it can be concluded that with an increase in the number of subjects and objects in the planning algorithm, there is a fairly strong increase in resource consumption, which complicates the process of cognitive evaluation of the complex situation for a large number of stakeholders. Despite the fact of the length and laboriousness of calculations, the pre-constructed action plan at times facilitates the tactical operation of the agent, providing possible auxiliary actions of other agents. A sing approach to planning provides an opportunity to preserve the experience accumulated in the process of acting agents, greatly accelerating the process of finding plans for solving similar problems which becomes obvious from the data of 7 and 8 experiments. Experiment 7 uses a task with similar objects, agents, and constraints with task 1, but the initial situation has been partially changed. In the 8th experiment, the experience of the agents is used in finding the plan in the first problem for finding the plans in the general problem with the 7th experiment.

6 Conclusion

The paper presents an algorithm for planning behavior by an intelligent agent with based on signs world model and functioning in a group of heterogeneous agents. The agent is given the task of achieving a certain goal in the space of conditions of the environment, which he is unable to achieve independently. In the sign based world model agents are described by signs "*I*" and "*Other*" ("*They*"), which allow them to include both their actions in the plan, and the actions of others. The proposed MultiMAP algorithm was tested on the model task "*Blocksworld*". In the future, it is planned to enable agents to assess the possible costs of performing actions and choose between different, both group and individual plans, for the least use of the resource. Also, to compile and select more complex plans, special sign protocols of communication between agents will be developed.

Acknowledgments. This work was supported by the Russian Science Foundation (Project No. 16-11-00048).

References

1. Brafman, R.I.: A privacy preserving algorithm for multi-agent planning and search. In: Proceedings of the Twenty-Fourth International Joint Conference on Artificial Intelligence (IJCAI 2015), pp. 1530–1536 (2015)
2. Emel'yanov, S., Makarov, D., Panov, A.I., Yakovlev, K.: Multilayer cognitive architecture for UAV control. Cogn. Syst. Res. **39**, 58–72 (2016)
3. Gerevini, A.E., Long, D.: Plan constraints and preferences in PDDL3. Technical report (2005)
4. Gupta, N., Nau, D.S.: On the complexity of blocks-world planning. Artif. Intell. **56**(2–3), 223–254 (1992)
5. Hayes, B., Scassellati, B.: Effective robot teammate behaviors for supporting sequential manipulation tasks. In: International Conference on Intelligent Robots and Systems, pp. 6374–6380 (2015)
6. Hélie, S., Sun, R.: Autonomous learning in psychologically-oriented cognitive architectures: a survey. New Ideas Psychol. **34**(1), 37–55 (2014)
7. Hexmoor, H., Eluru, S., Sabaa, H.: Plan sharing: showcasing coordinated UAV formation flight. Informatica (Ljubljana) **30**(2), 183–192 (2006)
8. Ivanov, D., Kalyaev, I., Kapustyan, S.: Formation task in a group of quadrotors. In: Kim, J.-H., Yang, W., Jo, J., Sincak, P., Myung, H. (eds.) Robot Intelligence Technology and Applications 3. AISC, vol. 345, pp. 183–191. Springer, Cham (2015). doi:10.1007/978-3-319-16841-8_18
9. Kahneman, D.: Thinking Fast and Slow. Penguin, New York (2011)
10. Kitano, H., Asada, M., Kuniyoshi, Y., Noda, I., Osawai, E., Matsubara, H.: RoboCup: a challenge problem for AI and robotics. In: Kitano, H. (ed.) RoboCup 1997. LNCS, vol. 1395, pp. 1–19. Springer, Heidelberg (1998). doi:10.1007/3-540-64473-3_46
11. Leontyev, A.N.: The Development of Mind. Erythros Press and Media, Kettering (2009)

12. Maliah, S., Shani, G., Stern, R.: Collaborative privacy preserving multi-agent planning. Auton. Agents Multi-Agent Syst. **31**(3), 493–530 (2017)
13. Osipov, G.S.: Sign-based representation and word model of actor. In: Yager, R., Sgurev, V., Hadjiski, M., Jotsov, V. (eds.) 2016 IEEE 8th International Conference on Intelligent Systems (IS), pp. 22–26. IEEE (2016)
14. Osipov, G.S., Panov, A.I., Chudova, N.V.: Behavior control as a function of consciousness. I. World model and goal setting. J. Comput. Syst. Sci. Int. **53**(4), 517–529 (2014)
15. Osipov, G.S., Panov, A.I., Chudova, N.V.: Behavior control as a function of consciousness. II. Synthesis of a behavior plan. J. Comput. Syst. Sci. Int. **54**(6), 882–896 (2015)
16. Osipov, G.S.: Signs-based vs. symbolic models. In: Sidorov, G., Galicia-Haro, S.N. (eds.) MICAI 2015. LNCS, vol. 9413, pp. 3–11. Springer, Cham (2015). doi:10.1007/978-3-319-27060-9_1
17. Panov, A.I.: Behavior planning of intelligent agent with sign world model. Biol. Inspired Cogn. Archit. **19**, 21–31 (2017)
18. Panov, A.I., Yakovlev, K.: Behavior and path planning for the coalition of cognitive robots in smart relocation tasks. In: Kim, J.-H., Karray, F., Jo, J., Sincak, P., Myung, H. (eds.) Robot Intelligence Technology and Applications 4. AISC, vol. 447, pp. 3–20. Springer, Cham (2017). doi:10.1007/978-3-319-31293-4_1
19. Panov, A.I., Yakovlev, K.S.: Psychologically inspired planning method for smart relocation task. Proc. Comput. Sci. **88**, 115–124 (2016)
20. Richter, S., Westphal, M.: The LAMA planner: guiding cost-based anytime planning with landmarks. J. Artif. Intell. Res. **39**, 127–177 (2010)

Using Augmentative and Alternative Communication for Human-Robot Interaction During Maintaining Habitability of a Lunar Base

Boris Kryuchkov[1], Leonid Syrkin[2], Vitaliy Usov[1],
Denis Ivanko[3,4(✉)], and Dmitriy Ivanko[4]

[1] Yu. Gagarin Research and Test Cosmonaut Training Center,
Star City, Moscow Region, Russia
{b.kryuchkov,v.usov}@gctc.ru
[2] State Region Social-Humanitarian University,
Kolomna, Moscow Region, Russia
syrkinld@mail.ru
[3] St. Petersburg Institute for Informatics and Automation of the Russian
Academy of Sciences (SPIIRAS), St. Petersburg, Russia
denis.ivankoll@gmail.com
[4] ITMO University, St. Petersburg, Russia

Abstract. The experience of the ISS missions has demonstrated that it is necessary to examine the potential of using robots assistants for maintaining a habitability of the space station and for improving the quality of life of cosmonauts in lunar missions. New manned space missions will be marked by a decrease in crew support from Earth. These conditions call for the new forms of Human-Robot Interaction (HRI) in order to manage the workload of crew members in context of using robots for maintaining habitability of the manned space complexes. Correct setting of the daily schedule of robots activities in space missions will contribute to decrease the operational time of cosmonauts for routine manipulations, including Housekeeping manipulations, searching and preparing means for Next Day's Work, transfer cargo, food preparation, elimination of waste and others works. An Augmentative and Alternative Communication (AAC) is a simple and clear variant of the HRI, designed to explain the meaning of the task for a robot. AAC is widely used for disabled persons with impairment in ability for communication skills, but in case of design HRI these techniques may be intended for the specification of tasks for robots on base of Ontology for Robotics. In accordance with AAC approach every schematic encodes for HRI may include the schedule time for job and its contents (or tools of necessity), as well as the conditions for initiating the activity or terminating it. This form of communication in the context of not-too-high intellectual capabilities of a robot (as a communication partner) may be considered as a practical approach in typical situations for fulfilling scenarios of daily activities in accordance with the 24-h schedules of the crew.

Keywords: Human Robot Interaction (HRI) · Augmentative and Alternative Communication (AAC) · Robot assistant · Lunar base's habitability

© Springer International Publishing AG 2017
A. Ronzhin et al. (Eds.): ICR 2017, LNAI 10459, pp. 95–104, 2017.
DOI: 10.1007/978-3-319-66471-2_11

1 Introduction

When building the HRI for use in space missions we must take into account the bio-social nature of human, since in such conditions the unity of biological and social life manifests itself very clear. We can specify the following distinctions from usual conditions on Earth: a cosmonaut works onboard within a small working group (with other members of space crew), he is constantly lives in artificial environment, he has to constantly monitor the status of onboard control systems of the spacecraft and the status of life-support systems [6]. Missions cost is exceptionally expensive, so we intend to obtain the more valuable output from cosmonauts activities, but not at the expense of the duration of the works, as it will lead to fatigue and reduced performance, but from using robot-assistants for indoors housekeeping. First of all, robots will be used to meet the physiological needs of people: the organization of nutrition, recreation, water consumption, of hygienic procedures, etc. In space there are difficulties with preservation of biorhythms of a human body, and for these purposes we need to apply the "Sleep–Wake" cycle, including the building of scenarios for robots activity within definite periods of time.

The activity of a cosmonaut in flights is unthinkable without its organization with the help of a daily schedule that includes blocks with intended use of the "biological functions", of works with payload and maintenance of onboard systems, performing space experiments and physical training for the prevention of negative factors of flight, and sessions of communication. A considerable amount of time are allocated for prep work, cleaning, storing items and tools, that is, job to maintain the habitability of the station in the broadest sense of the term. It seems likely that robots, intended for indoors activities, can help to reduce the work load on a cosmonaut when performing simple in composition manipulation works, such as for maintaining the habitability of the station or ensure physiological needs of crew members.

A variety of operations to maintain the habitability of the pressurized facility and the satisfaction of physiological needs of the crewmembers leads to the necessity to develop a system of communication that comes from the ontology of this subject area, but does not require excessive complexity of a human-machine interface [5]. This kind of compromise is possible when we use the ideas of "assistive support" of handicapped persons (HCP) in the field of conducting communications with a deficit of mental functions of perception, thinking, speech production etc.

It is obvious that semantic content of an upcoming action that is assigned for a robot has a strong cognitive component, which is the basis to analyze how such problems can be solved by using Augmentative and Alternative Communication (AAC) in the case of HCP. Inasmuch as these approaches received positive feedback from many experts, we will analyze the theoretical and methodological basis and propose specific types of information for conducting communication between humans and robots at the level of the semantic content of the current situation and the prescribed script of the upcoming action.

2 The Usage of Service Robots During Maintaining Habitability of a Lunar Base

The prospects of participation of robot-assistants in the maintenance of safe life conditions in space to ensure habitability for humans in extreme environments are in focus in a large number of publications [1, 3, 6].

There are objective difficulties and constrictions due to the extraordinary heterogeneity and diversity of tasks for which they are applicable. Some of these problems are possible to solve by constructing the Human – Robot Interaction (HRI) based on multimodal interfaces and by improvement the quality of communication between humans and robots [5, 11]. Perspective directions of using robot-assistants on manned space complexes are shown in Fig. 1.

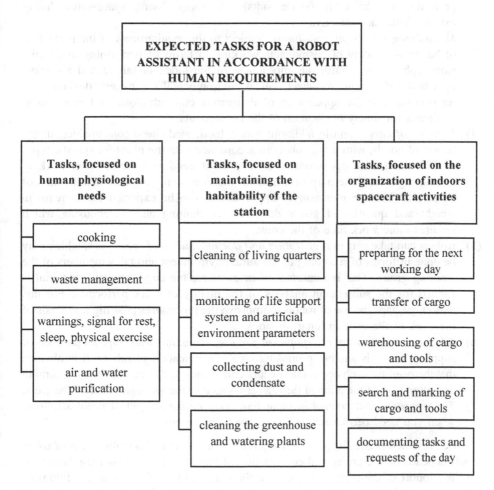

Fig. 1. Perspective directions of using robot-assistants to reduce the workload of cosmonauts on space complexes

Based on the composition of a typical "24-h timelines" crew schedule, it is possible to allocate priority directions of the use of a robot-assistant.

(1) *Information support.* Based on 24-h schedule of a crew, a robot can keep a record of the execution of the works, can signal a significant deviation from the time parameters, and can identify a conflict of interest failure. A robot can offer to substitute for performing some routine functions in case of "spare time" of a crew in the schedule.

(2) *Monitoring by the robot an environment status*, in particular of artificially generated air environment can contribute for ensuring sanitary and hygienic requirements. In the same vein, it is necessary to consider the task of cleaning of premises, disposal of biological waste, etc. Obviously, on the lunar surface a factor in the contamination of the working environment in lunar module may become more obvious due to the nature of the soil, polluting the means for protection of the body (space suits) and tools during spacewalks during extravehicular activity (Eva).

(3) *Maintaining safe living conditions.* Subject to the requirements of the protection of health and safety of cosmonauts working in terms of performing lunar missions, robots can control the appearance of the index of failure and abnormal operations of systems to detect critical environmental parameters, deviations in temperature and the appearance of dangerous concentrations of harmful contaminants for timely notification of the cosmonauts.

(4) *Delivery of cargo*, and in addition: water, food, medicines, consumables, distribution of goods, with using robots in accordance with the plans for specific types of 24-h timeline of operations, is an important trend to decrease spent working time in flight for these purposes. In lunar missions searching and identification of cargo may become increasingly important, as can be expected the increase in weight and quantity of goods delivered, accounting and warehousing which requires more work time of the crew.

(5) Robots can take part in *registration and documentation of crew work*, which may be done in forms of recording of the aural environment and talks members of the working group. That is important in cases where the target objective material is formed for the subsequent analysis of the quality of work performed. For this reason, perhaps the use of robots to improve the procedures post-flight analysis of the work of the crew in lunar missions.

(6) *Advanced preparation of equipment for use,* including warm-up, preparation of supplies, etc. – is another promising area of application of robots. It is obvious that the need for such operations occurs continuously; this can be time-consuming and often exceed the allotted time, as envisaged in the work schedule of the crew.

(7) *The environmental control* in lunar base may be the subject of robot activities, when you need permanent service of space greenhouse.

The presented above list is not exhaustive as the amount of possible areas of using robots depends on the projects of construction of lunar bases and equipping them with the life support of the crew and to ensure the survivability of plants in conditions of extreme external environment.

3 Application of Scenarios for a Robot-Assistant Activity Related to 24-h Schedules of Crew Work

Biosocial nature of human imposes the significant restrictions on the time and the appearance of activity for a robot to support the work of a small group, like a crew of a lunar mission. These restrictions are aimed at creating conditions for satisfaction the basic physiological needs of a person in the hermetically closed environment. In addition, the activity of a robot should not disturb the daily schedule of cosmonauts.

It should be noted, that nowadays for the ISS there is a highly developed and quite advanced system of planning which is to proactively prepare the month's work plan and applied it on the current day. Today it is difficult to say how existing approaches will be extended with move from orbital flights to lunar missions, where informational support of the Mission Control Center (MCC) will be reduced and delayed in time. There is reason to believe that this will reveal the trend to greater autonomy of the crew for operational planning based on prepared "templates" standard 24-h timelines for a flight day. In this case, a robot should not create additional interference with the daily living of the crew. This is perhaps the most important social role of robotic.

The development of a method of communication of cosmonauts with a robot-assistant must be based on the experience of the crew of the Russian segment (RS) of the ISS with the daily planning. This time-scale of the planning is interesting because the crew has a full understanding of the regulations of works, and sees the whole "outline" of events and to a certain extent can reproduce the fragment of the ontology that defines the relationships of the active agents (robots) with the objects of labor and tools to influence them in the context of the targets. If in a specific period of flights the responsible dynamic operations (docking with the transport ships, operations and the extravehicular activity (Eva), associated with increased security requirements) are not take place, then for a typical 24-h periods used so-called "form 24" for the specific calendar day. Its destination is to build the sequence of tasks for each individual member of the crew, and the whole crew if necessary. In fact, the same data type can be specified for the activity of a robot, which includes the degree of involvement of a robot in the activities of a particular member of the crew. But the base attributes must be signs of "permission - prohibition" for robot's activity with taken into consideration the parameters of localization in a particular module.

Let us briefly examine what blocks are in the schedule of works can be identified to make the planning activity of the robot is more clear and transparent.

In manned flights, cosmonauts try to keep a habitual cycle "sleep - wakefulness". A dream stands out for 8.5 h. The time of getting up usually happens at 6 am. Thus, a typical day during wakefulness cosmonauts are split into 6.5: average time on morning toilet up to half an hour, a few more to 40 min is given for Breakfast and about an hour for lunch. For prevention of negative influence of flight factors 2.0–2.5 h are planned for physical training for each crew member. The inspection of the onboard status takes a short time (about 10 min) with fixing several parameters of the life support system. Some time is issued for implementation to bond with MCC, planning the day, preparing for work (reading instructions and gathering tools), and familiarization with the work plan. And in addition a period of time spent preparing for sleep, hygiene, breakfast and dinner. Weekly cleaning lasts up to 3 h, usually is on a Saturday. Saturday and Sunday are days off.

These data may vary depending on the composition of the planned flight operations, but even this brief overview shows that in flight it is necessary to provide methods of planning of the robot activity and methods for operational tasks so that for all members it was convenient and comfortable.

It is necessary to specify the order of interaction between crew members when planning the activity of a robot-assistant to avoid conflicts of interest.

The potentially available planning scheme of robot-assistant use is shown in Fig. 2.

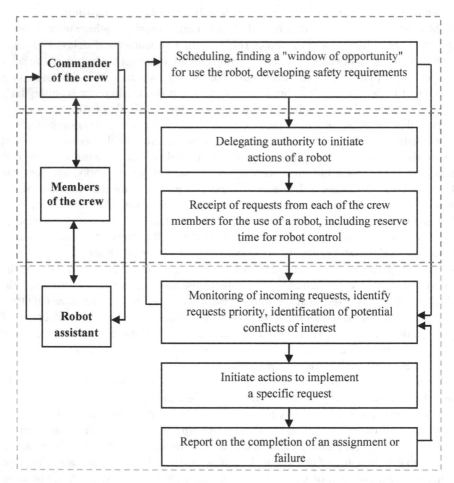

Fig. 2. Scheme of planning robot-assistant actions with accordance of the accepted order of interaction between the crew members

For better use of robots is proposed to use the 24-h timeline of the robot's activity, in which each time interval is indexed with the sign of the "permission - prohibition activity", in some cases with reference to the location at the station. An additional feature – the kind of activity can be marked with a specific icon corresponding to one of the types of activities that a robot can do in offline mode.

To conclude this section we formulate the answer to the question, how significant for future missions to examine the HRI in social aspects of the organization of work in small groups at a lunar base, if the basic phases of the Moon exploration are planned in the next 15–20 years? We can definitely say that today, all countries participating in the manned space flight, went on to stage full-scale simulation, for example, – In Russia the project "Mars-500" was the long-term simulation of a manned flight, and a project "MOON-WALK", prepared by the ESA, is now in active phase. This means, which support daily life activities of a crew, are considered as basic conditions in interplanetary missions.

4 Theoretical and Methodological Basis for Use "Image Schemes" for AAC

HRI in the light of biosocial natures of human makes the need for clarification of the role of cognitive components of communication that reveals in solving typical problems of the organization of life. One possible approach can be based on the concepts of "image schemes" [2, 7, 10].

The concept of imaginative schemes has a deep foundation in cognitive linguistics.

For example, Johnson offers a point of view, claiming that the "image-schema is an expression of such a schematic structure, which organizes our experience" [10].

Another source to represent the planned actions to the robot can be found in the works of Minsky; Schenk and Abelson [8, 10].

In the field of cognitive linguistics it is known to use along with the term "frame" also the terms "script", "script", "situational model", "cognitive model", the "stage", "prototype", "map" of Fillmore [7, 10].

In addition to the systematization of patterns of the interaction of members of a crew with a robot-assistant on the basis of structurally-logic schemes, or image-schema (in terms of models of linguistic analysis Johnson [10], which reflected in image-schemas the most characteristic or typical actions and operations), we should take into account the individual characteristics of the semantic space that are studied in the framework of psycho-semantics through the reconstruction of personal meanings and determine individual values, according to Petrenko [9], "categorical structure of consciousness." From the standpoint of psycho-semantics the each specific implementation of the scenarios of a cosmonaut activity is largely dictated by the specifics of the individual values (of images, verbal and non-verbal signs, value orientations), which impose a certain imprint on the perception of the situation, "individual style" perform mission tasks and communication. However, practical implementation of any system of interaction should ensure the implementation of a number of General principles.

The most important of them is following principles [11]:

– the unity of command, involving the preferred execution of commands, addressed to the robot-assistant, under the General management of the commander of the crew;

- the priority of safety, restricting the powers of crew members on the formation of tasks to perform certain types of work in certain periods or under certain flight conditions;
- the priority of the planned tasks, fixed by the calendar plan and the daily routine;
- the priority of tasks related to the functioning of the life support system and systems aimed at maintaining the resource capabilities of the human body.

5   Proposed Framework

We suggest an interactive mode of construction the dialogue of crew members with a robot during the organization of its operation and initialization of its tasks. We illustrate it through the description of the hypothetical scenario of building a dialogue with the robot.

(1) At the initial moment of time when a cosmonaut makes the decision about the need to connect the robot to carry out a specific work, he opens on a personal mobile device a dialog box and sees the state of the robot. This state can be represented by using the conditional image of the robot, painted in one of the following colors, that are represented green – readiness to receive the job, yellow is the execution mode of the previous job and the state of waiting its completion, red – no readiness to initiate a new action (loss of function, the prohibition of activity on the part of the commander of the crew, prohibition of activity according to a 24-h work schedule).

(2) If the image of the robot indicates its readiness to work, a cosmonaut gives a signal to initiate the script (special sign "attention") and from that moment the robot passes under the control of this particular member of the crew. Only the chief of the crew can cancel this interaction in case of emergency situations.

(3) Further we propose to carry out formation of an "image scheme" with help of the composition of a number of signs and images, which are subordinated to the logic of building a frame-like structure. Among the most important of its components will be included: (a) setting the time of the robot activity (this may be in the form of directly specifying the start time and end time, and may be an indication of the time of day, if it is not critical in terms of performance); It is significant that in any case the robot should follow the verification of the possibility of fulfilling the prescription, based on the limitations of the general schedule of the crew; if it is impossible to reconcile the specified time with a General schedule, the robot issues a conflict of interest message; (b) the type of the actions of the robot; if the stage of setting the time of work is successful, the next step is to set the type of actions from the existing list of categories (there should be few, each category must have its own encoding in the form of an icon or a symbol); (c) the location, where the robot is prescribed for activity (localization can be linked to an electronic map of the entire station, where the color shows the zones of acceptable and unacceptable activities in the context of flight operations, the operation mode of the onboard equipment and the work schedule of crew members).

(4) If we specify types of actions from the existing list of categories and there are several variants of the task, the robot puts the list of objects from which the cosmonaut must make a choice. For example, if the specified action is encoded with the sign of a trash collection ("panicle" or "trash" or some other image with the same semantic content) then a refinement could be "dust cleaning" or "condensate collection". In the first case can be used the image of a vacuum cleaner, and the second – "water drops".

It can be stated that in general case we are talking about using frame-like structures that allow building frames nested into each other, forming the description of the meaning of a course of action. This corresponds to the widespread practice approach to the concept of the frame when the frame is described as a structure of knowledge representation, or a tool of representation of cognitive structures, in this case of the way of action ("image-action"). That is, in this case the frame is considered as a method of knowledge representation, which represents a scheme of actions of the robot in a real situation. This definition is most closely to the interpretation "Frames for presentation of knowledge" of Minsky [8].

As indicated in [4], "this approach to build intelligent systems for high-level robot control (control system of a robot behavior) with use of frame-like structures may be a basis for designing a dialogue between a human and a robot".

6 Conclusions

This paper discusses specific applications of robots to maintain the habitability of the hermetically closed space object (for example, a lunar station) and to improve the conditions of staying inside a small working group. However, this approach can be extended to more general situations, where a small group works in a remote location, for example, on a marine drilling rig, on the Arctic base, etc. In addition, it was shown that for a case of a small group we should be especially attentive to the problems of role positions distribution in a team when using a robot-assistant. To this effect, in the composition of the knowledge bases of a robot (its model of the external world) it should be envisaged the criteria to organize incoming tasks from different people: not only in the queue of incoming requests from the group members, but also given the specified role positions, criteria of urgency and safety, etc.

Human-robot interaction (HRI) is the object of an interdisciplinary study because robot-assistants are poised to fill a growing number of roles in today's society [11, 12]. Nowadays robots are becoming involved in increasingly more complex and less structured tasks and activities in space missions, including interaction with cosmonauts required to complete those tasks [6]. This has prompted new problems how humans can interact with robots and accomplish interactive tasks in human environments, using the form of communication with a semantic context [2, 4, 6].

Assessing the prospects for further development of this approach it is possible to ascertain the need to pay more serious attention to the design of set of symbols with help of which it is supposed to convey the semantic content of the activity's pattern. For this purpose, the plan of the experimental studies in form of "double blind control" is proposed in which three groups of respondents will be involved: the first group will performs the semantic content of the activity's pattern, the second group will suggest the design of symbolic images with specific requirements for robot's job and the third group will try to recognize the prescribed actions. The expert group will evaluate the reasons for success or failure of this process, to identify systematic errors (human factors) or the role of personal factors. The experiment is expected to be held with the involvement of different age groups of respondents.

Acknowledgements. The research is partially supported by state research № 0073-2014-0005.

References

1. Bodkin, D.K., Escalera, P., Bocam, K.J.: A human lunar surface base and infrastructure solution, September 2006. doi:10.2514/6.2006-7336 (2006)
2. Falzon, P. (ed.): Cognitive Ergonomics: Understanding, Learning, and Designing Human-Computer Interaction. Academic Press, London (2015). 261 p.
3. Moroz, V.I., Huntress, V.T., Shevelev, I.L.: Planetnye ekspeditsii XX veka [Planetary expeditions of the XX century]. Space Res. **40**(5), 451–481 (2002). (in Russian)
4. Frame-like structures DynSoft. http://www.dynsoft.ru/contacts.php
5. Karpov, A., Ronzhin, A.: Information enquiry kiosk with multimodal user interface. Pattern Recogn. Image Anal. **19**(3), 546–558 (2009)
6. Kryuchkov, B.I., Usov, V.M.: The human-centered approach to the organization of joint activities of cosmonauts and an anthropomorphic robot-assistant on manned spacecrafts. Manned Spaceflight **3**(5), 42–57 (2012). (in Russian)
7. Lakoff, G.: Cognitive modeling. In: Women, Fire and Dangerous Things. Language and Intelligence. pp. 143–184 (1996) (in Russian). Women, Fire, and Dangerous Things: What Categories Reveal About the Mind (trans: George, L.). University of Chicago Press, Chicago (1987). ISBN 0 226 46804 6
8. Minsky, M.: Frames for presentation of knowledge. Energiya, Moscow, 151 p. (1979) (in Russian). A Framework for Representing Knowledge (trans: Minsky, M.). MIT-AI Laboratory Memo 306, June 1974. Winston, P. (eds.) The Psychology of Computer Vision. McGraw-Hill, New York (1975, reprinted) (1974)
9. Petrenko, V.F.: Introduction to experimental psychosemantic: a study of forms of representation in ordinary consciousness, Moscow, 177 p. (1983) (in Russian)
10. Pospelov, G.S.: Preface. In: Minskiy, M. (ed.) Frames for Presentation of Knowledge. Energiya, Moscow 151 p. (1978) (1987, in Russian)
11. Yusupov, R., Kryuchkov, B., Karpov, A., Ronzhin, A., Usov, V.: Possibility of application of multimodal interfaces on a manned space complex to maintain communication between cosmonauts and a mobile robotic assistant. Manned Spaceflight **3**(8), 23–34 (2013). (in Russian)
12. Kryuchkov, B.I., Karpov, A.A., Usov, V.M.: Promising approaches for the use of service robots in the domain of manned space exploration. SPIIRAS Proc. **32**(1), 125–151 (2014). doi:10.15622/sp.32.9. (in Russian)

An Application of Computer Vision Systems to Unmanned Aerial Vehicle Autolanding

Alexey Y. Aksenov, Sergey V. Kuleshov,
and Alexandra A. Zaytseva[(⊠)]

Saint-Petersburg Institute for Informatics and Automation of RAS,
14th Line V.O., 39, Saint-Petersburg 199178, Russia
{a_aksenov,kuleshov,cher}@iias.spb.su

Abstract. The paper considers an approach to autolanding system for multi rotor unmanned aerial vehicle based on computer vision and visual markers usage instead of global positioning and radio navigation systems. Different architectures of autolanding infrastructure are considered and requirements for key components of autolanding systems are formulated.

Keywords: Computer vision · Unmanned aerial vehicle · Autolanding

1 Introduction

The rapid development of technology of autonomous robotic systems and the need of unmanned aircraft in various fields, including military purposes, had led to the accelerated pace of development and deployment of small unmanned (UAVs) and remotely piloted aircraft. The most actively developing aircraft are the multi motor flying platforms (multicopters) [1, 2]. The common features for all objects of given class are the design and the principle of flight. The central part of multicopter consists of the control unit, battery and cargo holder. The micro engines with rotors are mounted radially from the centre on beams forming star-shaped layout of copter. To compensate the twist due to momentum the rotors rotate in different directions. Nevertheless such a symmetrical layout assumes presence of front and rear parts considering the direction of flight. During the flight the multicopter maintains relative to ground horizontal position, move sideways, change altitude and is able to hover. In presence of additional equipment the automatic and half-automatic flights are possible. To perform a movement the multicopter is needed to be taken out of balance by throttling combinations of rotors. As a result multicopter tilts and starts to fly in needed direction. To rotate multicopter clockwise the front and rear rotors are spining-up and left and right rotors are slowing down. The counterclockwise twist is done by analogy [3].

Multi motor UAVs have several advantages compared to other unmanned and manned aircraft. Unlike the helicopter the multi motor UAVs are less expensive to maintain, more stable in air, and are easier to control which result in higher ability to survey small ground objects with high spatial resolution [4].

In general multi motor UAV is a universal, efficient and simple type of unmanned vehicle which can take advantage over traditional helicopter design on the market and

© Springer International Publishing AG 2017
A. Ronzhin et al. (Eds.): ICR 2017, LNAI 10459, pp. 105–112, 2017.
DOI: 10.1007/978-3-319-66471-2_12

become useful instrument for mass media, photo and video industry etc. The greatest interest in UAVs show government agencies and services which functions are related to the protection, control and monitoring of the facilities [5].

2   Computer Vision Aided Autolanding System

The key element of autolanding system is an accurate spatial positioning of aircraft. For this task a number of subsystems could be used: internal inertial navigation systems, global positioning systems, external guidance (including computer vision (CV) systems) [6, 7].

To succeed in autolanding task the following requirements are to be met:

- unambiguous spatial positioning of aircraft including orientation, heading angle and altitude above landing site;
- functioning in wide range of weather and environment conditions;
- robustness of markers localization methods during spatial maneuvers of UAV;
- minimal time of aircraft position parameters obtaining.

This paper proposes an approach to autolanding system for multi rotor or helicopter UAVs based on computer vision without usage of global positioning systems and radio navigation.

Global positioning systems such as GPS/GLONASS and inertial navigation systems are efficient during flight phase but doesn't provide sufficient accuracy needed in landing phase leading to inability to solve the problem for small heights and areas [8, 9].

The typical accuracy of modern GPS receivers in horizontal plane reaches 6–8 m provided a good satellite visibility and correction algorithms. Over the territory of USA, Canada and limited number of other countries it is possible to increase accuracy to 1–2 m using the differential mode. Thus the use of GPS system over other territories appears to be inefficient.

The CV based system of automated landing [1, 10, 11] consists of the following components (Fig. 1):

- camera or camera array for locating of the guidance markers;
- system of markers suitable for efficient capturing by digital camera;
- image processing unit (CV-core) capable of real time obtaining of UAV's coordinates relative to landing site by processing digital camera image sequences;
- command unit computing landing command sequence for UAV's control core according to information from CV-core;
- data transmission channel.

The following options for system components deployment are available:

- the cameras are located above landing site which gives the ability to control both of landing zone and the UAV. The components of CV-core and command unit are located stationary on the ground (Fig. 2). As a variant of application that utilizes both markers and computer vision it is worth to mention a multicopter collective control project developed by Institute for Dynamic Systems and Control (IDSC) at ETH

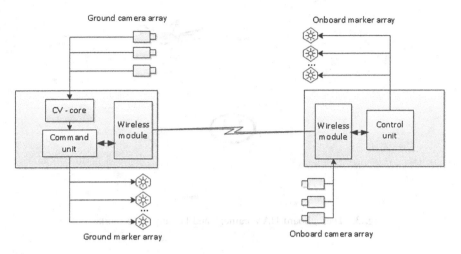

Ground camera array

Onboard marker array

Ground marker array

Onboard camera array

Fig. 1. CV based automated landing system

Zurich [12]. The computer vision system with two or four cameras is used to auto-matically maintain the given direction of flight. This system achieves high accuracy but needs preliminary prepared workspace because the markers are located on UAV and the computer vision system is located stationary outside the UAV [12, 13];

- cameras are located on UAV and observe the markers on landing zone (Fig. 3);
- cameras are located on landing site in upward direction observing the markers on UAV (Fig. 4);

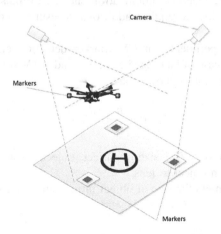

Camera

Markers

Markers

Fig. 2. External cameras above landing site

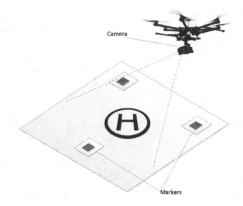

Fig. 3. The onboard UAV camera and landing zone markers

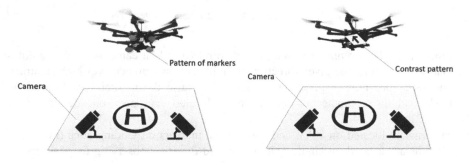

Fig. 4. Cameras are located on landing ground

The placement of cameras on UAV requires either onboard CV processing module either a low latency broadband communication channel to transmit images to ground CV processing unit. Besides it requires additional measures to decouple camera from UAV's rotors vibration.

Thus for further experiments on CV marker detection and UAV altitude and position assessment the case of landing site camera and UAV side markers was chosen.

From geometric optics it is known that:

$$1 + \frac{h}{H} = \frac{d}{f},$$

where h – linear object size, H – size of object projection on focus plane, f – focal length, d – distance from object to lens.

In case of digital camera the size of object projection on camera sensor determines as follows:

$$H = p\frac{L}{S},$$

where p – object projection size in pixels, L – linear size of camera sensor, S – photo sensor size in pixels.

Thus provided camera focal length and linear object size the distance from object to camera sensor can be formulated by following expression:

$$d = f\left(1 + \frac{hS}{pL}\right) = f\left(1 + \frac{hR}{p}\right),$$

where $R = S/L$ (pixels/mm) – pixel density of camera sensor.

Experimental evaluation of CV altitude measurement for the case of single camera on landing site and 4 LED-markers on UAV is shown on Fig. 5.

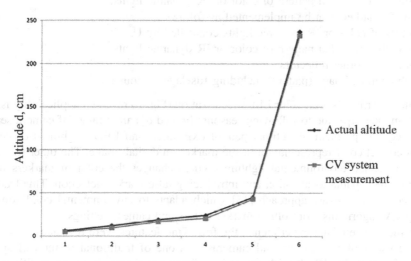

Fig. 5. Evaluation of UAV altitude measurement by CV autolanding system

The minimal operational altitude is determined by UAVs projection size does not exceed the size of sensor, which results in:

$$d_{min} = f\left(1 + \frac{hR}{\min(S_H, S_W)}\right),$$

where S_H, S_W – camera sensor horizontal and vertical definition in pixel respectfully.

The maximal operational altitude is determined by minimal discernible distance between UAV markers p_{min}:

$$d_{max} = f\left(1 + \frac{hR}{p_{min}}\right).$$

In practice it is difficult to obtain steady results with $p_{min} < 5$ pixels.

For example, given the camera with 7 mm focal length, 640 × 480 pixel resolution, 100 pixel/mm sensor density and UAV size of 55 mm the following operational parameters are obtained: $d_{min} = 87$ mm, $d_{max} \sim 7,7$ m.

3 Marker Usage

All of camera positioning options are demanding the usage of special markers to increase efficiency of CV-core algorithms [1, 14].

Landing site markers can be implemented as follows:

- directional pattern of high contrast images suitable for recognition by CV-core;
- remotely controlled pattern of color or IR dynamic lights;
- UAV's markers can be implemented as follows:
- pattern of color or IR dynamic lights controlled by UAV;
- remotely controlled pattern of color or IR dynamic lights;
- pattern of corner reflectors;
- high contrast image (pattern) including fuselage contour itself.

The experiments with color LEDs localization (Fig. 6) revealed applicability issues of this marker type due to following reasons: limited dynamic range of camera sensor introduces the problem of optimal pair of exposition and LED brightness choice to obtain correct color representation of the marker on digital image. The other reason is the uncontrolled environmental lightning which changes the color of markers interfering correct spatial localization and introducing false marker detection. This leads to more complex software approach [15] which adapts to environmental conditions by tuning CV algorithms, controlling marker colors and camera settings.

In the course of the experiments, the following features were revealed:

The usage of IR markers and cameras is a one of traditional method of spatial objects location [16, 17]. It's advantage is resistance to various lighting conditions and background changes, but there are some problems with the direction toward the sun giving a strong IR illumination. Besides it requires additional means to identify individual markers for UAV heading detection.

Fig. 6. Color LED markers location problem (a – original image, b – Blue LEDs localization, c – Red LEDs localization) (Color figure online)

The color markers (including RGB LEDs) are mainly applicable in controlled lightning (ideal) conditions, but uncontrolled environmental effects (fog, dust, colored lightning) introduces unreliable marker recognition making automatic impossible. Usage of controlled RGB lights neglect environmental effects to some extent but requires wireless communication channel with UAV.

High contrast geometry marker patterns demand good visibility and/or lightning conditions.

The conducted studies showed that the most universal and efficient are the dynamic light markers controlled by landing system computer via wireless communication channel. In this case computer turns on the UAV's markers independently and waits its appearance on images obtained from cameras. It simplifies the procedure of marker detection and localization which in turn makes possible to use timing factor in conjunction with prior knowledge of marker's state resulting in simple CV methods of detection implementation such as difference image. Besides, it doesn't require usage of markers with different properties to identify its position on UAV because it is sufficient to light only one specific at a time to identify it.

Nevertheless the mode of controlled lights markers requires more images to obtain spatial UAV's location due to sequential switching of all markers for detection. Each marker demands minimum of one image to detect it ON/OFF state. To overcome this problem the usage high frame rate cameras are required.

The controlled lights markers are preferable when it is possible to organize communication channel between UAV and automatic landing system.

4   Conclusion

The paper proposes an approach to development of automatic landing system for multi rotor UAVs based on computer vision instead of global positioning systems, functional diagram of the system is developed.

It is shown that despite of camera position the usage of markers simplifies the location of key points of UAVs for CV system and thus enhances obtaining the spatial position of aircraft.

The calculation of operational altitude limits is given for effective UAV markers capturing by CV system resulting in successful autolanding procedure.

The various options for markers placement and types were analyzed and following recommendations were developed:

The remote-controlled lights markers are always preferable when it is possible to organize communication channel between UAV and automatic landing system.

The color markers (including RGB LEDs) are mainly applicable in controlled lightning (ideal) conditions.

High contrast geometry marker patterns demand good visibility and/or lightning conditions.

The most universal and efficient are the dynamic light markers controlled by landing system computer via wireless communication channel. In this case computer turns on the UAV's markers independently and waits its appearance on images obtained from cameras. It simplifies the procedure of marker detection and localization which in turn makes possible to use timing factor in conjunction with prior knowledge of marker's state resulting in simple CV methods of detection implementation such as difference image. Besides, it doesn't require usage of markers with different properties to identify its position on UAV because it is sufficient to light only one specific at a time to identify it.

References

1. Aksenov, A.Y., Kuleshov, S.V., Zaytseva, A.A.: An application of computer vision systems to solve the problem of unmanned aerial vehicle control. Transp. Telecommun. **15**(3), 209–214 (2014)
2. Altug, E., Ostrowski, J.P., Taylor, C.J.: Control of a quadrotor helicopter using dual camera visual feedback. Int. Rob. Res. **24**(5), 329–341 (2005)
3. Schmid, K.: View planning for multi-view stereo 3D reconstruction using an autonomous multicopter. J. Intell. Rob. Syst. **65**(1–4), 309–323 (2012)
4. Barbasov, V.K.: Multirotor unmanned aerial vehicles and their capabilities for using in the field of earth remote sensing. Ingenernye izyskaniya **10**, 38–42 (2012). (in Russian)
5. Zinchenko, O.N.: Unmanned aerial vehicles: the use of aerial photography in order to map. P.1. Racurs, Moscow, 12 p. (2012) (in Russian)
6. Saripalli, S., Montgomery, J.F., Sukhatme, G.S.: Visually guided landing of an unmanned aerial vehicle. IEEE Trans. Rob. Autom. **19**(3), 371–380 (2003)
7. Garcia Carrillo, L.R., Dzul Lopez, A.E., Lozano, R.: Combining stereo vision and inertial navigation system for a quad-rotor UAV. J. Intell. Rob. Syst. **65**, 373 (2012). doi:10.1007/s10846-011-9571-7
8. Cesetti, A., Frontoni, E., Mancini, A., Zingaretti, P., Longhi, S.: A vision-based guidance system for UAV navigation and safe landing using natural landmarks. In: 2nd International Symposium on UAVs, Reno, Nevada, USA, pp. 233–257, 8–10 June 2009
9. Corke, P.: An inertial and visual sensing system for a small autonomous helicopter. J. Rob. Syst. **21**(2), 43–51 (2004)
10. Cesetti, A., Frontoni, E., Mancini, A.: A visual global positioning system for unmanned aerial vehicles used in photogrammetric applications. J. Intell. Rob. Syst. **61**, 157 (2011). doi:10.1007/s10846-010-9489-5
11. Levin, A., Szeliski, R.: Visual odometry and map correlation. In: IEEE Computer Society Conference on Computer Vision and Pattern Recognition, Washington, D.C., USA (2004)
12. ETH IDSC. Flying Machine Arena, Zurich (2014). http://www.idsc.ethz.ch
13. Ritz, R., Müller, M.W., Hehn, M., D'Andrea, R.: Cooperative quadrocopter ball throwing and catching. In: Proceedings of Intelligent Robots and Systems (IROS), IEEE/RSJ International Conference, Vilamoura, October 2012, pp. 4972–4978. IEEE (2012)
14. Open Computer Vision. http://sourceforge.net/projects/opencvlibrary/. Accessed May 2017
15. Kuleshov, S.V., Yusupov, R.M.: Is softwarization the way to import substitution? SPIIRAS Proc. **46**(3), 5–13 (2016). doi:10.15622/sp.46.1. (in Russian)
16. Kuleshov, S.V., Zaytseva, A.A.: The selection and localization of semantic frames. Inf. J. Inf.-Measur. Control Syst. **10**(6), 88–90 (2008). (in Russian)
17. Kuleshov, S.V., Zaytseva, A.A.: Object localization of semantic blocks on bitmap images. SPIIRAS Proc. **7**, 41–47 (2008). (in Russian)

About Approach of the Transactions Flow to Poisson One in Robot Control Systems

Eugene Larkin[1(✉)], Alexey Bogomolov[1], Dmitriy Gorbachev[1], and Alexander Privalov[2]

[1] Tula State University, Tula 300012, Russia
elarkin@mail.ru
[2] Tula State Pedagogical University, Tula 300026, Russia
privalov.61@mail.ru

Abstract. Flows of transactions in digital control systems of robots are investigated. On the base of the fact, that uses the conception of Poisson character of transactions flow permits to simplify analytical simulation of robot control process, the problem of estimation the degree of approach of real flow to Poisson one is putted on. Proposed the criterion based on evaluation of expectation of waiting function. On the example of investigation of "competition" in the swarm of robots it is shown that flow of transactions, generated by swarm, when quantity of robots aspire to infinity approximately aspire to Poisson one.

Keywords: Robot digital control system · Transaction · Poisson flow · Non-poisson flow · Regression · Correlation · Parametric criterion · Pearson's criterion · Waiting function

1 Introduction

Functioning of mobile robots may be considered as sequence of switching from one state of onboard equipment to another under control of commands flow [1, 2]. States of robot may include interpretation of the programs by onboard computer [3, 4], receiving/generating transactions, execution the commands by mechanical or electronic units, service a queue of command [5, 6], support the dialogue with remote operator [7] etc. Below, all switches and interconnections will be called transactions. Transactions occur in the physical time. Time intervals between transactions for the external observer are random values [8]. So transactions form a flow. One of variety of flows is the stationary Poisson flow, which possesses an important property – absence of aftereffect [9, 10]. Use such a property permits substantially to simplify the modeling behavior of robots. So when working out robot's digital control systems there is always emerges a question about a degree of approximation the real flow to Poisson flow. Methods of estimation of properties the flows of transactions are developed insufficiently, that explains the necessity and relevance of the work.

© Springer International Publishing AG 2017
A. Ronzhin et al. (Eds.): ICR 2017, LNAI 10459, pp. 113–122, 2017.
DOI: 10.1007/978-3-319-66471-2_13

2 Quantitative Estimations of Approximation Degree of Transaction Flows to Poisson One

It is well known, that time intervals between transactions in Poisson flow are characterized with exponential distribution law [9–11]:

$$f(t) = \frac{1}{T}\exp\left(-\frac{t}{T}\right), \tag{1}$$

where T – is the expectation of time interval; t – is the time.

Regression criterion is based on estimation of standard-mean-square error as follows [12, 13]:

$$\varepsilon_r = \int_0^\infty [g(t) - f(t)]^2 dt, \tag{2}$$

where $g(t)$ – is the distribution under estimation.

Let $g(t) = \delta(t - T)$, where $\delta(t - T)$ – is Dirac δ-function. Then:

$$\varepsilon_r = \int_0^\infty [\delta(t - T) - f(t)]^2 dt = \varepsilon_{r1} + \varepsilon_{r2} + \varepsilon_{r3}, \tag{3}$$

where

$$\varepsilon_{r1} = \int_0^\infty \delta^2(t - T)dt = \lim_{a \to 0} \int_{T-a}^{T+a} \left(\frac{1}{2a}\right)^2 dt = \infty;$$

$$\varepsilon_{r2} = -2\int_0^\infty \delta(t - T_g) \cdot \frac{1}{T_f}\exp\left(-\frac{t}{T}\right)dt = -\frac{2}{eT};$$

$$\varepsilon_{r3} = \int_0^\infty \frac{1}{T^2}\exp\left(-\frac{2t}{T}\right)dt = \frac{1}{2T}.$$

Thus criterion ε_r changes from 0 (flow without aftereffect) till ∞ (flow with deterministic link between transactions), and it has the dimension as [time−1].

Correlation criterion is as follows [14]:

$$\varepsilon_c = \int_0^\infty g(t) \cdot \frac{1}{T}\exp\left(-\frac{t}{T}\right)dt. \tag{4}$$

This criterion changes from $\frac{1}{2T}$ (flow without aftereffect) till $\frac{1}{eT}$, where $e = 2,718$ (deterministic flow). With use the function:

$$\tilde{\varepsilon}_c = \frac{e(1 - 2T\varepsilon_c)}{e - 2}. \tag{5}$$

Criterion may be done the non-dimensional one, and it fits the interval $0 \leq \tilde{\varepsilon}_c \leq 1$. Parametrical criterion is based on the next property of exponential low (1) [9–11]:

$$T = \sqrt{D}, \tag{6}$$

where $D = \int_0^\infty \frac{(t-T)^2}{T} \exp\left(-\frac{t}{T}\right) dt$ – is the dispersion of the law (1).

To obtain from the property (6) non-dimensional criterion, fitting the interval $0 \leq \tilde{\varepsilon}_c \leq 1$, one should calculate the next function:

$$\varepsilon_p = \frac{(T_g - \sqrt{D_g})^2}{T_g^2}, \tag{7}$$

where $T_g = \int_0^\infty tg(t)dt$ and $D_g = \int_0^\infty (t - T_g)g(t)dt$ – are expectation and dispersion of density $g(t)$.

In the case of experimental determining $g(t)$ as a histogram:

$$g(t) = \begin{pmatrix} t_0 \leq t < t_1 & \cdots & t_{i-1} \leq t < t_k & \cdots & t_{J-1} \leq t < t_J \\ n_1 & & n_i & & n_J \end{pmatrix}, \tag{8}$$

where n_i – is quantity of results fitting the interval $t_{i-1} \leq t < t_i$, then estimation of proximity of $f(t)$ and $g(t)$ one should use Pearson's criterion [15–17].

3 Criterion, Based on "Competition" Analysis

Let us consider transactions generation process as the "competition" of two subjects: external observer and transaction generator. Model of "competition" is the 2-parallel semi-Markov process, shown on the Fig. 1a.:

$$\mathbf{M} = [A, \mathbf{h}(t)], \tag{9}$$

where $A = \{a_{w1}, a_{w2}, a_{g1}, a_{g2}, \}$ – is the set of states; a_{w1}, a_{g1} – are the starting states; a_{w2}, a_{g2} – are the absorbing states; $\mathbf{h}(t)$ – is the semi-Markov matrix:

$$\mathbf{h}(t) = \begin{bmatrix} \begin{bmatrix} 0 & w(t) \\ 0 & 0 \end{bmatrix} & \mathbf{0} \\ \mathbf{0} & \begin{bmatrix} 0 & g(t) \\ 0 & 0 \end{bmatrix} \end{bmatrix}; \quad \mathbf{0} = \begin{bmatrix} 0 & 0 \\ 0 & 0 \end{bmatrix}. \tag{10}$$

116　　E. Larkin et al.

Let us consider the situation, when observer "wins" and at from the moment τ waits the event, when occurs the next transaction. For determine the waiting time let us construct on 2-parallel process M the ordinary semi-Markov process M' (Fig. 1b):

$$M' = [A', h'(t)],\qquad(11)$$

where $A' = A \cup B$ – is the set of states; $A = \{\alpha_1, \alpha_2, \alpha_3\}$ – is the subset of states, which simulate beginning and ending of wandering through semi-Markov process; α_1 – is the starting state; α_2 – is the absorbing state, which simulate "winning" of transaction generator; α_3 – is the absorbing state, which simulates end of waiting by the observer the event of generation transaction; $B = \{\beta_1, \ldots, \beta_i, \ldots\}$ – is the infinite set of states, which define time intervals for various situation of finishing of transaction generator; $h'(t) = \left\{ h'_{m,n}(t) \right\}$ – semi-Markov matrix, which define time intervals.

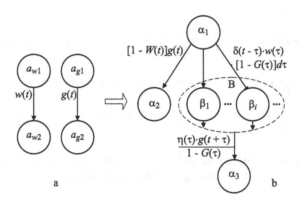

Fig. 1. To waiting time calculation

Elements $h'_{m,n}(t)$ are as follows: $h'_{m,n}(t)$ is the weighted density of time of finishing transaction generator if it "wins" the "competition":

$$h'_{12}(t) = g(t)[1 - W(t)],\qquad(12)$$

where $W(t) = \int_0^t w(\theta)d\theta$ – is the distribution function; θ – is an auxiliary variable;

$h'_{1,2+i}(t)$, $i = 1, 2, \ldots$, are defined as weighted densities of time of finishing of observer exactly at the time τ, if he "wins" the "competition" and waits the transaction:

$$h'_{1,2+i}(t) = \delta(t - \tau) \cdot w(\tau)[1 - G(\tau)]d\tau,\qquad(13)$$

where $\delta(t - \tau)$ – Dirac function; $G(t) = \int_0^t g(\theta)d\theta$; $w(\tau)[1 - G(\tau)]d\tau$ – probability of finishing the observer exactly at the time τ, if he "wins" the "competition";

$\frac{\eta(t)\cdot g(t+\tau)}{1-G(\tau)}$ – is the density of time of residence semi-Markov process $\mathbf{h}'(t)$ in state B, where $\eta(t)$ – is the Heaviside function.

Thus, probability of hitting the process $\mathbf{h}'(t)$ to the state B is as follows $p_{\alpha_0\beta} = \int\limits_0^\infty [1 - G(\tau)]w(\tau)d\tau = \int\limits_0^\infty W(t)g(t)dt$. Weighted density of time of waiting by observer the next transaction is equal to $h_{w\to g}(t) = \eta(t)\int\limits_0^\infty w(\tau)g(t+\tau)d\tau$. Pure density is as follows:

$$f_{w\to g}(t) = \frac{\eta(t)\int\limits_0^\infty w(\tau)g(t+\tau)d\tau}{\int\limits_0^\infty W(t)dG(t)}. \tag{14}$$

Let us consider function $f_{w\to g}(t)$ behavior when $g(t) = \frac{1}{T_g}\exp\left(-\frac{t}{T_g}\right)$ (flow of transactions without aftereffect) and when $g(t) = \delta(t - T_g)$ (flow with deterministic link between transactions).

Formula (14) for the first case is as follows:

$$f_{w\to g}(t) = \frac{\eta(t)\int\limits_0^\infty w(\tau)\frac{1}{T_g}\exp\left[-\frac{t+\tau}{T_g}\right]d\tau}{1 - \int\limits_{t=0}^\infty \left[1 - \exp\left(-\frac{t}{T_g}\right)\right]dW(t)} = \frac{1}{T_g}\exp\left(-\frac{t}{T_g}\right). \tag{15}$$

Formula (14) for the second case is as follows:

$$f_{w\to g}(t) = \frac{\eta(t)w(T_g - t)}{W(T_g)}. \tag{16}$$

Suppose that $w(t)$ have range of definition $T_{w\,min} \le arg\,w(t) \le T_{w\,max}$ and expectation $T_{w\,min} \le T_w \le T_{w\,max}$. In dependence of location $w(t)$ and $g(t)$ onto time axis, it is possible next cases:

(a) $T_g < T_{wmin}$. In this case (14) is senseless.

(b) $T_{wmin} \le T_g \le T_{wmax}$. In this case $f_{w\to g}(t)$ is defined as (16), range of definition is $0 \le arg[f_{w\to g}(t)] \le T_g - T_{wmin}$, and $\int\limits_0^\infty tf_{w\to g}(t)dt \le T_g$.

(c) $T_g > T_{wmax}$. In this case $f_{w\to g}(t) = w(T_g - t)$, $T_g - T_{wmax} \le arg[f_{w\to g}(t)] \le T_g - T_{wmin}$, and $\int\limits_0^\infty tf_{w\to g}(t)dt \le T_g$.

For this case density of waiting time by Dirac function when transaction occurs is as follows:

$$f_{\delta \to g}(t) = \frac{\eta(t) \cdot g(t + T_g)}{\int\limits_T^\infty g(t)dt}. \tag{17}$$

Expectation of $f_{\delta \to g}(t)$ is as follows:

$$T_{\delta \to g} = \int\limits_0^\infty t \frac{g(t + T_g)}{\int\limits_T^\infty g(t)dt} dt. \tag{18}$$

So, the criterion, based on "competition" analysis, is as follows:

$$\varepsilon_w = \left(\frac{T_g - T_{\delta \to g}}{T_g} \right)^2, \tag{19}$$

where T_g – is the expectation of density $g(t) = \delta(t - T_g)$; $T_{\delta \to g}$ – is the expectation of density $f_{\delta \to g}(t)$. For the exponential law:

$$\varepsilon_w = \left(\frac{T - T_{\delta \to g}}{T} \right)^2 = \left(\frac{T - T}{T} \right)^2 = 0. \tag{20}$$

Let us investigate behavior of $T_{\delta \to g}$. For that let us expectation of $g(t)$ as (Fig. 2):

$$\int\limits_0^\infty tg(t)dt = \int\limits_0^T tg(t)dt + \int\limits_0^\infty tg(t + T_g)dt + T_g \int\limits_0^\infty g(t + T_g)dt \tag{21}$$

$$= p_{1g}T_{1g} + p_{2g}T_{\delta \to g} + p_{2g}T_g = T_g,$$

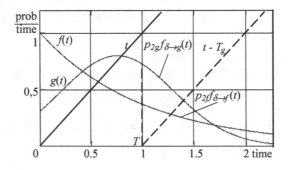

Fig. 2. To calculate of expectation

where $p_{1g} = \int\limits_0^T g(t)dt; \quad p_{2g} = \int\limits_T^\infty g(t)dt.$

If $g(t) = f(t)$, then $p_{1f} = \frac{e-1}{e}; \; p_{2f} = \frac{1}{e}; \; T_{1f} = T\frac{e-2}{e-1}$ and from equation:

$$p_{1f}T_{1f} + p_{2f}T_{\delta \to f} + p_{2f}T = T, \tag{22}$$

follows that:

$$T_{\delta \to f} = T. \tag{23}$$

If $g(t) \neq f(t)$ then from (21) follows that:

$$T_{\delta \to g} = \frac{p_{1g}(T - T_{1g})}{1 - p_{1g}}. \tag{24}$$

4 Example

As an example let us consider the case, when transaction flow is formed as a result of competition inside the swarm of robots, when common quantity of robots is equal to K. Time from the start of observation and transaction, formed by k-th robot, $1 \leq k \leq K$, is defined with identical uniform lows. Forming the flow of transactions may be simulate with K-parallel semi-Markov process:

$$\mathbf{M}^K = [\mathbf{A}^K, \mathbf{h}^K(t)], \tag{25}$$

where $A^K = \{a_{11}, \ldots, a_{k1}, \ldots, a_{K1}, a_{12}, \ldots, a_{k2}, \ldots, a_{K2}\}$ – is the set of states; $a_{11}, \ldots, a_{k1}, \ldots, a_{K1}$ – is the subset of starting states; $a_{12}, \ldots, a_{k2}, \ldots, a_{K2}$ – is the subset of absorbing states; $\mathbf{h}^K(t)$ – semi-Markov matrix, kk-th elements of the main diagonal of which is equal to $\begin{bmatrix} 0 & v_k(t) \\ 0 & 0 \end{bmatrix}$, and other elements are equal to zeros (Fig. 3):

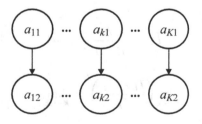

Fig. 3. Forming a flow of transactions in swarm of K robots

$$v_1(t) = \ldots = v_k(t) = \ldots v_K(t) = v(t) = \begin{cases} 1, & when\ 0 \le t \le 1; \\ 0\ in\ all\ other\ cases. \end{cases} \quad (26)$$

K-parallel process starts from all states of subset $a_{11}, \ldots, a_{k1}, \ldots, a_{K1}$ contemporaneously. Transaction is generated when one of ordinary semi-Markov processes gets the state from the subset $a_{12}, \ldots, a_{k2}, \ldots, a_{K2}$. In accordance with theorem by Grigelionis [18] when $K \to \infty$ flow of transactions approaches to Poisson one.

Density of time when at least one of robot of swarm generates transaction is as follows:

$$g_K(t) = \frac{d\{1 - [1 - V(t)]^K\}}{dt}, \quad (27)$$

where:

$$V(t) = \int_0^t v(\tau)d\tau = \begin{cases} 2t, & when\ 0 \le t \le 1; \\ 0\ in\ all\ other\ cases. \end{cases}$$

For this case:

$$g_K = \begin{cases} K(1-t)^{K-1}, & when\ 0 \le t \le 1; \\ 0\ in\ all\ other\ cases. \end{cases} \quad (28)$$

Expectation of (27) is as follows:

$$T_K = \int_0^1 tK(1-t)^{K-1}dt = \frac{1}{K+1}[time]. \quad (29)$$

Exponential law, which define Poisson flow of transactions is as follows:

$$f_K(t) = (K+1)\exp[-(K+1)t]\left[\frac{prob}{time}\right]. \quad (30)$$

For waiting function:

$$\tilde{T}_K = \frac{K}{(K+1)^2}[time], \quad (31)$$

$$\lim_{K \to \infty} \varepsilon_{lg}^K = \lim_{K \to \infty}\left(\frac{T_K - \tilde{T}_K}{T_K}\right)^2 = \lim_{K \to \infty}\frac{1}{(K+1)^2} = 0. \quad (32)$$

I.e. with increasing K law approaches to exponential, in accordance with B. Grigelionis theorem. Densities are shown on the Fig. 4. Already at $K = 0,012$ what can be considered as a good approximation.

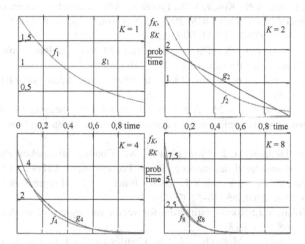

Fig. 4. Densities of time between transactions from swarm of robots

5   Conclusion

In such a way criteria with which may be estimated degree of approach of transaction flow in robotic system to Poisson flow were investigated. Criterion, based on calculation of the waiting time, is proposed, as one, which gives good representation about properties of flow of transactions, and have respectively low runtime when estimation.

Further investigation in the domain may be directed to description of method of calculations when processing statistics of time interval in a flow. Other research may concern the errors, to which leads the modeling of robot with non-Poisson character of transactions flows, when Poisson flows conception was laid into the base of simulation.

References

1. Tzafestas, S.G.: Introduction to Mobile Robot Control. Elsevier, Amsterdam (2014)
2. Kahar1, S., Sulaiman1, R., Prabuwono1, A.S., Akma, N., Ahmad, S.A., Abu Hassan, M.A.: A review of wireless technology usage for mobile robot controller. In: International Conference on System Engineering and Modeling (ICSEM 2012), International Proceedings of Computer Science and Information Technology IPCSIT, vol. 34, pp. 7–12 (2012)
3. Thrun, S., Burgard, W., Fox, D.: Probabilistic Robotics. Intelligent Robotics and Autonomous Agents series. MIT Press, Cambridge (2005)
4. Ching, W.-K., Huang, X., Ng, M.K., Siu, T.K.: Markov Chains: Models, Algorithms and Applications. International Series in Operations Research & Management Science, vol. 189. Springer Science+Business Media New York, New York (2013)

5. Pang, G., Zhou, Y.: Two-parameter process limits for an infinite-server queue with arrival dependent service times. Stoch. Process. Appl. **127**(5), 1375–1416 (2017)
6. Mathijsen, B., Zwart, B.: Transient error approximation in a Lévy queue. Queueing Syst. **85** (3), 269–304 (2017)
7. Larkin, E.V., Ivutin, A.N., Kotov, V.V., Privalov, A.N.: Interactive generator of commands. In: Tan, Y., Shi, Y., Li, L. (eds.) ICSI 2016. LNCS, vol. 9713, pp. 601–609. Springer, Cham (2016). doi:10.1007/978-3-319-41009-8_65
8. Limnios, N., Swishchuk, A.: Discrete-time semi-Markov random evolutions and their applications. Adv. Appl. Probab. **45**(1), 214–240 (2013)
9. Markov, A.A.: Extension of the law of large numbers to dependent quantities. Izvestiia Fiz.-Matem **2**, 135–156 (1906)
10. Lu, H., Pang, G., Mandjes, M.: A functional central limit theorem for Markov additive arrival processes and its applications to queueing systems. Queueing Syst. **84**(3), 381–406 (2016)
11. Bielecki, T.R., Jakubowski, J., Niewęgłowski, M.: Conditional Markov chains: properties, construction and structured dependence. Stoch. Process. Appl. **127**(4), 1125–1170 (2017)
12. Maronna, R.M., Victor, J., Yohai, V.J.: Robust functional linear regression based on splines. Comput. Stat. Data Anal. **65**, 46–55 (2013)
13. Muler, N., Yohai, V.J.: Robust estimation for vector autoregressive models. Comput. Stat. Data Anal. **65**, 68–79 (2013)
14. Rank, M.K.: Correlation Methods, vol. 196. Charles Griffin & Company, London (1955)
15. Boos, D.D., Stefanski, L.A.: Essential Statistical Inference: Theory and Methods, vol. 568 (XVII). Springer, New York (2013)
16. Wang, B., Wertelecki, W.: Density estimation for data with rounding errors. Comput. Stat. Data Anal. **65**, 4–12 (2013)
17. Luedicke, J., Bernacchia, A.: Self-consistent density estimation. Stata J. **14**(2), 237–258 (2014)
18. Grigelionis, B.: On the convergence of sums of random step processes to a Poisson process. Theory Probab. **8**, 177–182 (1963)

Modified Spline-Based Navigation: Guaranteed Safety for Obstacle Avoidance

Roman Lavrenov[1(✉)], Fumitoshi Matsuno[2], and Evgeni Magid[1]

[1] Kazan Federal University, Kazan 420008, Russian Federation
{lavrenov,magid}@it.kfu.ru
[2] Kyoto University, Kyoto 615-8540, Japan
matsuno@me.kyoto-u.ac.jp

Abstract. Successful interactive collaboration with a human demands mobile robots to have an advanced level of autonomy, which basic requirements include social interaction, real time path planning and navigation in dynamic environment. For mobile robot path planning, potential function based methods provide classical yet powerful solutions. They are characterized with reactive local obstacle avoidance and implementation simplicity, but suffer from navigation function local minima. In this paper we propose a modification of our original spline-based path planning algorithm, which consists of two levels of planning. At the first level, Voronoi-based approach provides a number sub-optimal paths in different homotopic groups. At the second, these paths are optimized in an iterative manner with regard to selected criteria weights. A new safety criterion is integrated into both levels of path planning to guarantee path safety, while further optimization of a safe path relatively to other criteria is secondary. The modified algorithm was implemented in Matlab environment and demonstrated significant advantages over the original algorithm.

Keywords: Path planning · Safety · Potential field · Voronoi graph

1 Introduction

Contemporary robotic applications target for human replacement in diverse scenarios that spread from social-oriented human-robot interaction [11] and collaboration [13] to automatic swarm control [12] and urban search and rescue in hostile environments [9]. All such applications demand indoor and outdoor autonomous path planning and navigation abilities with simultaneous localization and mapping (SLAM) [2], collaboration with other robots [10] and other functionality.

Path planning distinguishes global and local approaches. While the later operates in a completely unknown environment and robots make immediate decisions that are based on locally available information only, the global approach can access complete knowledge about environment, i.e., robot shape, initial and

© Springer International Publishing AG 2017
A. Ronzhin et al. (Eds.): ICR 2017, LNAI 10459, pp. 123–133, 2017.
DOI: 10.1007/978-3-319-66471-2_14

goal postures, and a set of environment obstacles are known in advance. Within a global approach model, potential field based methods [14], [4] could provide a globally defined potential function such that a goal position is represented as an attractive pole and obstacles are represented with repulsive surfaces. Next, a robot follows the potential function gradient toward its minimum [1]. Two main problems of potential field methods are oscillations in narrow passages between obstacles and a failure to select a good global potential function, which in turn results in local minima issues.

Our previous research had proposed a global approach path planning spline-based algorithm for a car-like mobile robot [7]. It uses potential field for obstacle avoidance and provides a locally sub-optimal path with regard to path length, smoothness and safety optimization criteria. In this paper we propose a modification of our original algorithm, which targets to improve robot safety with regard to environment obstacles. The path planning is performed in two stages: first, Voronoi-based approach provides a number sub-optimal paths in different homotopic groups; next, one of these paths is optimized in an iterative manner with regard to selected criteria weights. The algorithm integrates safety criteria into both levels of path planning in order to guarantee path safety, while further optimization of a safe path relatively to other path evaluation criteria is secondary. The new algorithm was implemented in Matlab environment and demonstrated significant advantages over the original algorithm.

The rest of the paper is organized as follows. Section 2 briefly describes our previous research. Section 3 presents a new criterion that improves the algorithm performance with regard to robot safety, and our modified spline-based algorithm, which successfully overcomes the weaknesses of the initial approach and uses advantages of the new optimization criterion. Section 4 compares the results of the original and the new algorithm, demonstrating the superiority of the later. Finally, we conclude in Sect. 5.

2   Original Spline-Based Navigation Algorithm Drawbacks

In our previous research [7] we had proposed a spline-based method that navigates an omnidirectional circle-shape robot in a 2D configuration space with known a-priori static obstacles. Each obstacle is approximated with a finite set of intersecting circles of different size. Given a start and a target positions, the robot searches for a collision-free path being guided by a cost function with pre-determined weights of each criteria.

Collision avoidance is managed with a repulsive potential function with a high value inside of a circular obstacle and a small value in free space. High values of the function push points of a path outside obstacles to minimize path cost during optimization. The potential field begins to decrease drastically on obstacle boundary and wanes rapidly with distance while moving away from the boundary. A contribution of a single circular obstacle repulsive potential at robot position $q(t) = (x(t), y(t))$ in time t is described with the following equation:

$$U_{rep}(q) = 1 + \tanh(\alpha(\rho - \sqrt{(x(t) - x)^2 + (y(t) - y)^2})) \quad (1)$$

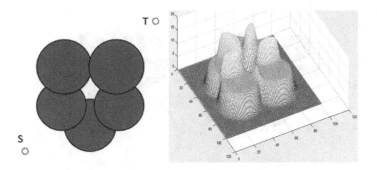

Fig. 1. Example of environment with five obstacles and start/target points (left) and repulsive potential function of Eq. 1 for $\alpha = 0.5$ (right) that corresponds to the obstacles

where ρ is the radius of the obstacle with centre (x, y) and α is an empirically defined parameter that is responsible for pushing a path outside of an obstacle. Figure 1 (right) demonstrates the example with $\alpha = 0.5$ for a single obstacle that is formed by five intersecting circles (Fig. 1, left), and the potential function has undesirable peaks at the circle intersections. Next, influence of all N obstacles of the environment are accumulated into *topology* $T(q)$ parametric function that is defined within $[0, 1]$:

$$T(q) = \sum_{j=0}^{N-1} \int_{t=0}^{1} U_{rep}^{j}(q) \cdot \delta l(t) \cdot dt \qquad (2)$$

where $\delta l(t)$ is a segment length. Smoothness property of the path is considered through *roughness* $R(q)$ function and is also integrated along the path:

$$R(q) = \sqrt{\int_{t=0}^{1} (x''(t))^2 + (y''(t))^2 dt} \qquad (3)$$

And the path length $L(q)$ sums up the lengths of all path segments:

$$L(q) = \int_{t=0}^{1} \delta l(t) \cdot dt \qquad (4)$$

The final path cost function sums up the three components with empirically predefined weights $\gamma_{i=1...3}$:

$$F(q) = \gamma_1 T(q) + \gamma_2 R(q) + \gamma_3 L(q) \qquad (5)$$

For example, obstacle penalty influence component is defined as $\gamma_1 = \frac{\beta}{2}$, where β ranges over an array that correlates with array of α parameters from Eq. 1 [6].

The original algorithm works iteratively, starting from a straight line (between S and T points) initial path. This line forms a first spline that utilizes three points: S, T and an equidistant point in between. Equation 5 sets the

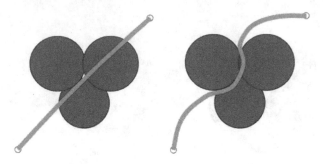

Fig. 2. Compound obstacle: first iteration (left) and final path after 18 iterations (right)

path cost, which is further optimized with Nelder-Mead Simplex Method [5] to minimize the total cost. A resulting *better* path serves as an initial guess for the next iteration. The optimization deals only with the points which define the spline, while path evaluation accounts for all points of the path. The spline is rebuilt at each iteration using information from a previous stage, increasing the number of spline's points by one. The algorithm terminates when number of iterations overflow a user-defined limit or if a new iteration fails to improve a previous one.

The original method is successful for simple obstacles that could be approximated with a single circle, while for compound obstacles a good selection of initial path becomes essential. Compound obstacles that consist of intersecting circles introduce potential field local maxima at each intersection (e.g., Fig. 1, right), which may trap a path in cost function local minimum as a next iteration spline can not overcome local maxima due to a local nature of the optimization process. Figure 2 demonstrates an example with one compound obstacle; three intersecting circles produce local repulsive potentials that sum up in a such way that an optimization succeeds to avoid local maxima, but stops after 18 iterations and reports a failure of providing a collision free path (Fig. 2, right).

3   Voronoi Graph Based Solution with Integrated Safety

We propose a modification of our original spline-based path planning algorithm for a mobile robot. The new algorithm consists of two levels of path planning. At the first level, Voronoi-based approach provides a number sub-optimal paths in different homotopic groups. At the second level, one of these paths is optimized in an iterative manner with regard to selected criteria weights. The algorithm integrates safety criteria into both levels of path planning in order to guarantee path safety, while further optimization of a safe path relatively to other path evaluation criteria is secondary.

3.1   Minimal Distance from Obstacles Criterion

First, we introduce an additional path quality evaluation criterion - minimal distance from obstacles. While navigating in obstacle populated environments, a robot should maximize its distance from the obstacles, and this feature is integrated into T(q) component of our cost function. However, in hostile environments, there may be an explicit requirement to stay at least at some minimal distance from obstacle boundaries - these may include semi-structured debris of urban scene, which risk to collapse further, or a danger of short-range emissions from the obstacles, etc. In such cases, the use of some paths that may be optimal with regard to Eq. 5 should be forbidden due to violation of minimal distance requirement, and a different path should be selected. A minimal distance of a robot from all obstacles of environment is calculated in each configuration $q(t)$ along the parametrically defined path as follows:

$$m(C) = \min_{\forall t \in [0,1]} dist_c(q(t)) \tag{6}$$

where $dist_c(q(t))$ is a minimal distance from obstacles in configuration $q(t)$:

$$dist_c(q(t)) = \min_{\forall c \in C} \sqrt{(x(t) - x(c))^2 + (y(t) - y(c))^2} - r(c) \tag{7}$$

Here C is a set of all circular obstacles c with the centre at $(x(c), y(c))$ and radius $r(c)$; further, these elementary circular obstacles may intersect to form compound obstacles. Also the user is required to specify a particular minimally acceptable distance to the boundaries of any obstacle d_m, and them optimization criterion D(q) is defined as follows:

$$D(q) = \omega^{d_m - m(C)} \tag{8}$$

where ω is a sufficiently large empirically defined value that prevents a path from approaching obstacles beyond the permitted limit. Value of this function is one when minimal distance to obstacles $m(C)$ is equal to safe value d_m, but it gains large positive value when the limit is violated, and diminishes to a small positive value when the robot keeps safe distance from an obstacle. We combine all four criteria within the total cost function for the optimization procedure as follows:

$$F(q) = \gamma_1 T(q) + \gamma_2 R(q) + \gamma_3 L(q) + \gamma_4 D(q) \tag{9}$$

where γ_4 is the weight for minimum distance criteria influence. Figure 3 demonstrates the criterion influence on the path: while with $\gamma_4 = 0$ it does not contribute to the total cost (left sub-figure), with $\gamma_4 = 1$, $d_m = 3$ and $\omega = 20$ the path avoids touching the obstacles (right sub-figure); other parameters are defined as $\gamma_1 = 1$, $\gamma_2 = 1$, and $\gamma_3 = 0.5$ for the both cases.

3.2   Selecting a Path with Voronoi Graph

In order to provide a good initial spline that could be further improved locally with regard to user selection of the cost weights, we apply Voronoi Diagram

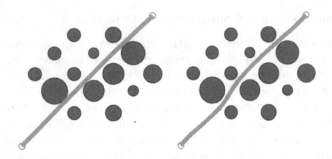

Fig. 3. The path moves away from obstacles when we use the new criterion

approach [15]. This helps to avoid the drawbacks of original algorithm, which is caused by a poor selection of an initial spline. The previously mentioned first level of path planning includes three stages: preparing environment, constructing Voronoi graph VG that spans the environment, and selecting a number of sub-optimal paths within VG with regard to a single evaluation criterion from Eq. 9.

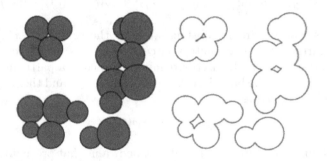

Fig. 4. Compound obstacles (left) and their external contours (right)

To prepare the environment, two steps are performed. First step groups intersecting circles $c \in C$ together in order to form a single compound obstacle O_1 through iterative growth. It continues to form such obstacles $O_i, i = 1 \ldots k$, where k is a number of compound obstacle within the environment, until all circles of the environment will be assigned to a particular compound obstacle. For example, there are four obstacles that are formed by groups of circles in Fig. 4 and three in Fig. 5. The second step shapes outer boundaries of each compound obstacle O_i of set $Obst = \{O_1, O_2, \ldots, O_k\}$. For example, this successfully removes three tiny internal contours inside compound obstacles in Fig. 4.

Next, Voronoi graph VG is constructed based on a classical brushfire approach [3]. Figure 5 (the middle image) demonstrates Voronoi graph VG construction example for the environment in Fig. 5(left). Thick blue lines are compound obstacles borders and thin blue lines depict emerging outwards and inwards rays. The rays intersection points are equidistant to nearest obstacles, and all together form VG, which is depicted with a thin red line in Fig. 5 (middle and right). At this stage we apply safe value d_m in order to remove from VG all edges that already do violate the safety requirements. Such example is demonstrated in Fig. 7 and will be explained in more details in the next section.

Fig. 5. Environment with obstacles (left), Voronoi graph building procedure (middle), the obtained Voronoi graph and the path within the graph (right) (Color figure online)

Finally, upon obtaining Voronoi graph VG, we finalize it by adding start S and target T nodes and perform a search of several sub-optimal paths within VG with Dijkstra algorithm; they are depicted with thick red lines in Fig. 5 (right). Next, a set of spanning points is extracted from this spline candidate and these points are utilized as via points for initial spline of the spline-based method [7]. As the path optimization with regard to Eq. 9 is performed only locally, the influence of additional parameter is also local. At the second level of path planning, a selected at the first level path is optimized in an iterative manner with a help of Eq. 9. More technical details about graph construction and spanning points selection could be found in [8].

4   Simulation Results

In order to verify our approach the new algorithm was implemented in Matlab environment and an exhaustive set of simulations was run. Particular attention was paid to the cases where the original algorithm fails [7]. The cost function of Eq. 9 was applied with empirical parameter selection $\gamma_1 = 1$, $\gamma_2 = 1$, $\gamma_3 = 0.5$, $\gamma_4 = 1$, $d_m = 3$ and $\omega = 20$. The algorithm succeeded to provide collision-free paths in all cases, which was a natural consequence of applying initial Voronoi-based path as an input for iterative optimization algorithm.

Figure 6 demonstrates environment, where the original spline-based algorithm had failed. At the first level, Voronoi graph VG provides us with a safe path without obstacle collisions. It serves as an initial path at the second level, which ensures a final path calculation with the modified spline-based algorithm

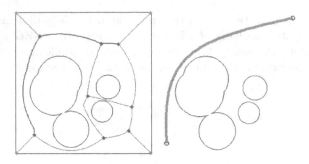

Fig. 6. Voronoi graph path (left) and corresponding spline-based optimal path (right)

within a significantly smaller number of iterations. Even though potentially significant time complexity of VG construction was not considered as an issue due to its off-line construction, the simulations empirically demonstrated that VG calculations take acceptably small amount of time for simple cases, while more simulations in complicated large-size environments are scheduled for the future.

For example, for the environment in Fig. 6 the Voronoi-based initial path calculation took only 2 s. The total running time of the new algorithm decreased in three times in average with regard to the original algorithm. This way, the final path in Fig. 6 was calculated in just 2 iterations within 2.5 min in Matlab, while the original spline-based algorithm had spent 18 iterations and 38 min to report its failure to provide a collision free path. Similarly, the original algorithm required 9 iterations and 15 min to provide a good path within Fig. 3 environment, while the new algorithm required 4 iterations and 5 min. In Fig. 4 the original algorithm failed to find a path, while the new algorithm successfully completed the task within 3 iteration and 2 min.

Voronoi graph (VG) contains multiple homotopy class paths and their variety depends on map complexity: the more distinct obstacles appear within the map, the larger is the amount of homotopy classes. In our simulated experiments we decided to limit the algorithm to no more than 5–7 homotopies. Next, these selected homotopies served as initial spline for a new smart spline-based method. We have tested this strategy with a number of environments. For each selected homotopy we verify minimal distance value d_m within a corresponding equidistant to obstacles edges and reject homotopies that pass in dangerous proximity of obstacles. The calculation of all suitable homotopies takes 1 to 3 min in average, depending on environment complexity. In Fig. 7 for $d_m = 3$, safe VD edges are depicted with green and forbidden edges in red colour (Fig. 7, left). Thus, only the homotopies that contain green edges could be considered for safe path candidates. After the first level of planning provided a candidate path, second level optimal path (Fig. 7, right) calculation took just 3 iteration and 2 min.

Figure 8 demonstrates an example where original spline-based method spends 32 min and 18 iterations before reporting a failure. The new algorithm calculates

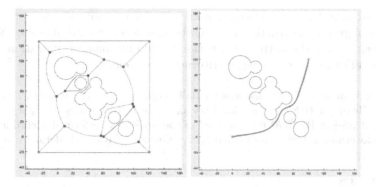

Fig. 7. Voronoi graph with dangerous edges (left) and a resulting safe path (right) (Color figure online)

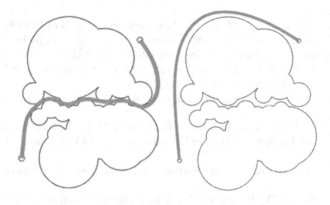

Fig. 8. Path with (right) and without minimal distance criterion (left)

5 homotopies within 3 min (Fig. 8 demonstrates paths in 2 different homotopy classes) and bases its selection on safety criterion (Fig. 8, right).

5   Conclusions

In this paper we presented a modification of our original spline-based path planning algorithm, which consists of two levels of planning. At the first level, Voronoi-based approach provides a number sub-optimal paths in different homotopic groups. At the second, these paths are optimized in an iterative manner with regard to selected criteria weights. A new safety criterion is integrated into both levels of path planning to guarantee path safety, while further optimization of a safe path relatively to other criteria is secondary. The modified algorithm was implemented in Matlab environment and demonstrated significant advantages over the original algorithm, including guaranteed path acquisition, successful avoidance of local minima problem, increased speed and guaranteed path safety in static planar environment. As a part of our future work, the algorithm will be tested in large-size environments in order to verify the acceptability of the Voronoi graph construction time at the first level of path planning. We also consider extending the path cost function with additional optimization criteria and perform exhaustive testing of criteria weight selection.

Acknowledgments. This work was partially supported by the Russian Foundation for Basic Research (RFBR) and Ministry of Science Technology & Space State of Israel (joint project ID 15-57-06010). Part of the work was performed according to the Russian Government Program of Competitive Growth of Kazan Federal University.

References

1. Andrews, J.R., Hogan, N.: Impedance control as a framework for implementing obstacle avoidance in a manipulator. M. I. T., Department of Mechanical Engineering (1983)
2. Buyval, A., Afanasyev, I., Magid, E.: Comparative analysis of ROS-based monocular SLAM methods for indoor navigation. In: Proceedings of SPIE, 9th International Conference on Machine Vision, vol. 10341, pp. 103411K–103411K-6 (2016)
3. Choset, H.M.: Principles of Robot Motion: Theory, Algorithms, and Implementation. MIT Press, Cambridge (2005)
4. Khatib, O., Siciliano, B.: Springer Handbook of Robotics. Springer, Heidelberg (2016)
5. Lagarias, J.C., Reeds, J.A., Wright, M.H., Wright, P.E.: Convergence properties of the Nelder-Mead simplex method in low dimensions. SIAM J. Optim. **9**(1), 112–147 (1998)
6. Magid, E.: Sensor-based robot navigation. Technion - Israel Institute of Technology (2006)
7. Magid, E., Keren, D., Rivlin, E., Yavneh, I.: Spline-based robot navigation. In: IEEE/RSJ International Conference on Intelligent Robots and Systems, pp. 2296–2301 (2006)
8. Magid, E., Lavrenov, R., Khasianov, A.: Modified spline-based path planning for autonomous ground vehicle. In: Proceedings of International Conference on Informatics in Control, Automation and Robotics (to appear)
9. Magid, E., Tsubouchi, T., Koyanagi, E., Yoshida, T.: Building a search tree for a pilot system of a rescue search robot in a discretized random step environment. J. Robot. Mechatron. **23**(4), 567–581 (2011)
10. Panov, A.I., Yakovlev, K.: Behavior and path planning for the coalition of cognitive robots in smart relocation tasks. In: Kim, J.-H., Karray, F., Jo, J., Sincak, P., Myung, H. (eds.) Robot Intelligence Technology and Applications 4. AISC, vol. 447, pp. 3–20. Springer, Cham (2017). doi:10.1007/978-3-319-31293-4_1
11. Pipe, A.G., Dailami, F., Melhuish, C.: Crucial challenges and groundbreaking opportunities for advanced HRI. In: IEEE/SICE International Symposium on System Integration, pp. 12–15 (2014)
12. Ronzhin, A., Vatamaniuk, I., Pavluk, N.: Automatic control of robotic swarm during convex shape generation. In: International Conference and Exposition on Electrical and Power Engineering, pp. 675–680 (2016)
13. Rosenfeld, A., Agmon, N., Maksimov, O., Azaria, A., Kraus, S.: Intelligent agent supporting human-multi-robot team collaboration. In: International Conference on Artificial Intelligence, pp. 1902–1908 (2015)

14. Tang, L., Dian, S., Gu, G., Zhou, K., Wang, S., Feng, X.: A novel potential field method for obstacle avoidance and path planning of mobile robot. In: IEEE International Conference on Computer Science and Information Technology, vol. 9, pp. 633–637 (2010)
15. Toth, C.D., O'Rourke, J., Goodman, J.E.: Handbook of Discrete and Computational Geometry. CRC Press, Boca Raton (2004)

Integration of Corporate Electronic Services into a Smart Space Using Temporal Logic of Actions

Dmitriy Levonevskiy[✉], Irina Vatamaniuk, and Anton Saveliev

St. Petersburg Institute for Informatics and Automation of the Russian Academy
of Sciences, 14th Line V.O. 39, 199178 St. Petersburg, Russia
DLewonewski.8781@gmail.com, vatamaniuk@iias.spb.su,
antoni-fox@yandex.ru

Abstract. This paper discusses integration of corporate information infrastructure components to develop an enterprise smart space that uses available resources to achieve the common goals of users. The paper proposes formalization for the smart space functioning scenarios on the basis of the temporal logic of actions, describes a multilevel model of corporate service integration and discusses the architecture of the Multimodal Information and Navigation Cloud System (MINOS).

Keywords: Smart space · Cyber-physical systems · Multimodal interfaces · Temporal logic of action · System integration

1 Introduction

Nowadays corporate information systems are actively being developed. They correspond to the Industry 4.0 trend, use cloud computing technologies, multimodal interfaces, cyber-physical systems (CPS) and adopt the intelligent environment approach (Smart Space) [1, 2]. These technologies can be used not only to increase interactivity, functionality and efficiency of the corporate environment and human-computer interaction (HCI), but also to make use of the possibilities of advanced technologies in various aspects of enterprise security, such as identification of emergency situations, regime violations, equipment failure. This can include human action recognition and classification [3], abnormal activity detection [4] and behavior forecasting [5].

Moreover, multimodal means of communication offered by cyber-physical systems significantly exceed the possibilities of traditional assistance systems. They enable communicating between user and system by means of gestures, voice, movements, etc. This fact simplifies interaction for untrained people [6] and for disabled users [7].

This paper discusses integration of corporate information infrastructure components to develop an enterprise smart space that uses available resources to achieve the common goals of the enterprise, for example, employee informing, providing security, event forecasting [8]. The interaction between users and components of the system is described by formalized scenarios tracked by the analytic subsystem that generates events according to user and system behavior.

© Springer International Publishing AG 2017
A. Ronzhin et al. (Eds.): ICR 2017, LNAI 10459, pp. 134–143, 2017.
DOI: 10.1007/978-3-319-66471-2_15

2 Problem Statement

Among the most serious problems that appear, while developing any automated systems is that the cost of development and support grows sharply with the growth of system complexity. Such systems are often produced as unique instances, and consequently cannot be cheap [9]. The complexity of Industry 4.0 systems makes unprofitable to develop them from scratch, so it is a common approach to apply the system integration methods. This works, but causes several problems.

Firstly, heterogeneous components of the enterprise infrastructure use different standards and technologies for communication, such as Bluetooth, Wi-Fi, UPnP, etc. The inconsistency of standards and technologies significantly complicates even basic actions (for example, transferring messages), especially when it comes to components manufactured by different vendors or working on different platforms [10].

Secondly, large amount of information gathered by remote sources and processed by various consumers aggravates problems of privacy, security and trust. Langeheinrich [11] describes a set of features that are specific for smart spaces and critical for security: ubiquity, invisibility, sensing and memory amplification. The problem of personal security is that it is difficult for users to comprehend data collecting and processing techniques [12] and to be aware of the amount and nature of gathered data, and for the system to control the data transfer. Another issue is related to the corporate security: intensive use of smart spaces in the enterprise activities increases the dependence on the information infrastructure, so the enterprise should take appropriate steps to protect it from attacks and ensure fault tolerance.

Thirdly, it should be noted that the actual possibilities of frameworks for smart space development are insufficient and the software debugging is rather complicated.

Thus, the general task is to integrate electronic services into a corporate smart space that will meet the open system requirements (portability, scalability, interoperability, etc.) and consist of items that possibly (and most likely) do not meet these requirements and were not designed to constitute the smart space. The components should be connected through standard network technologies. Moreover, reliability and security aspects of storing, processing and transferring data should be taken into consideration. Thus, it is necessary to synthesize the corporate smart space using methods of system integration, models of their interconnection, and related interfaces, protocols, algorithms and software architectures.

This work in particular considers building the logic-based model of human-computer interconnection scenarios used to track, analyze and forecast various behavior situations. It also proposes distributed application architecture and states the directions and issues of its development.

3 Scenario Models

Let us consider the example of the smart space being developed in the St. Petersburg Institute for Informatics and Automation of Russian Academy of Sciences (SPIIRAS). It includes the multimodal information and navigation cloud system (MINOS), some scenarios of its applications are shown on Fig. 1 [13].

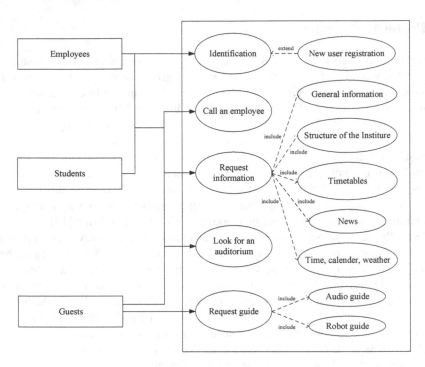

Fig. 1. Use case model for some interaction scenarios

The system interacts with user through the entry points (menus, screens, QR codes) that offer a set of services depending on the context and the user class. New users should get a system account and provide their biometric data. Guests of the institute can request the system for general institute information, look for an auditorium, call an employee by means of the videoconferencing service. In this case, a scenario can be represented as a graph. For example, Fig. 2 illustrates a generic user identification scenario: the subservices (nodes of the graph) constituting the identification service and the generated events (edges). The object-oriented data model of the identification service includes the components shown on Fig. 3.

On Fig. 3 for each identification method (Method) any user (Guest, Employee, Cooperator) may have a profile (Profile) containing ID, password, biometrics or other data that allow distinguishing him/her from other users. The user can be identified in zones (Zone) where the necessary equipment is installed (turnstiles, cameras, doors with electronic locks). The identification zones include external zones, entry and exit zones located on the border of the enterprise and in the external environment, and the internal ones located on the enterprise territory. Zones are included into enterprise areas (Area), i.e. floors, departments, etc. Facts of identification (Attempt) are stored in the database.

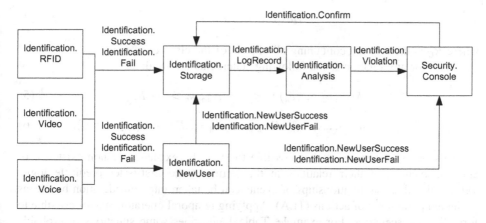

Fig. 2. User identification scenario

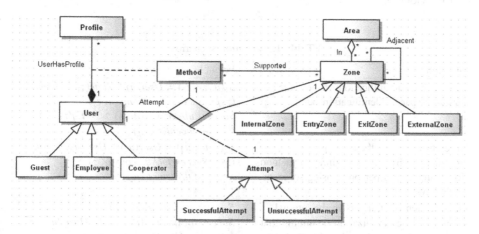

Fig. 3. UML class diagram for the user identification service

Class instances can be represented as entities:

$$e = <k, A>, \tag{1}$$

where k is the primary key of an entity, $A = <a_1, a_2, ..., a_n>$ is the set of its attributes. The entities with the same key are treated as equal:

$$Eq(e_1, e_2) : k_1 = k_2. \tag{2}$$

Let the following statements express that the given entity e belongs to the class C and that its attribute a is equal to v:

$$Is(e, C) : A \in R_C, \tag{3}$$

$$Has(e, a, v) : a = v, \tag{4}$$

where R_C is the relation containing all objects of the class C.

Also, for any relation R the following statements are used:

$$R(e_1, e_2, \ldots, e_n) : \; <k_1, k_2, \ldots, k_n> \; \in R, \tag{5}$$

$$R_{i1,i2,\ldots,im}(e_{i1}, e_{i2}, \ldots, e_{im}) : Ee_{im+1} Ee_{im+2} \ldots Ee_{in} R(e_1, e_2, \ldots, e_n). \tag{6}$$

This set of statements makes possible to state the existence or non-existence of entities, events and their relations in the terms of the first-order predicate logic. Temporal and causal relationships of events can be taken into consideration by means of the temporal logic of actions (TLA). Applying temporal operators makes possible to formalize the scenarios. For example, Table 1 describes some situations of probable security violations.

Table 1. Violation scenarios

Situation	Formula
Simultaneous identification of the same person at spaced identification points	$S_1(u, z_1, z_2) = User(u) \wedge Zone(z_1) \wedge Zone(z_2) \wedge SuccessfulAttempt(u, z_1) \wedge SuccessfulAttempt(u, z_2) \wedge \sim Eq(z_1, z_2) \wedge \sim Adjacent(z_1, z_2)$
Different results of the same person identification by means of different methods (for example, if a person is using someone else's keys)	$S_2(z, u_1, u_2) = User(u_1) \wedge User(u_2) \wedge Zone(z) \wedge SuccessfulAttempt(u_1, z) \wedge SuccessfulAttempt(u_2, z)$
Identification of strangers in the internal segment of the organization (assuming that any new visitors are introduced to the system at the first visit)	$S_3(u, z) = InternalZone(z) \wedge SuccessfulAttempt(u, z) \wedge Ap(Profile(p) \rightarrow \sim UserHasProfile(u, p))$
Identification of the person on the inside point of identification in the absence of entry mark	$S_4(u) = (Az_0 \; Entry(z_0) \rightarrow \sim SuccessfulAttempt(u, z_0)) \; U \; Ez (InternalZone(z) \wedge SuccessfulAttempt(u, z))$
Double entry	$S_5(u) = Ez_1 Ez_2 User(u) \wedge EntryZone(z_1) \wedge EntryZone(z_2) \wedge (SuccessfulAttempt(u, z_1) \; R \; (Az_3 \; ExitZone(z_3) \rightarrow \sim (SuccessfulAttempt(u, z_3)) \; R \; SuccessfulAttempt(u, z_2)))$

For any formula $S_i(x_1, x_2, \ldots, x_n)$ one can build logical statements about the presence (absence) of the situation i:

$$Ex_1 Ex_2 \ldots Ex_n S_i(x_1, x_2, \ldots, x_n), \tag{7}$$

$$Ax1 \, Ax2 \ldots Axn \sim Si(x1, x2, \ldots, xn). \tag{8}$$

The statements concerning more complex situations (for example, combination of S_i and S_j) can be built the same way. Applying the known methods of verification [14–16] allows determining whether the statements are true. That enables managing the event registry and taking actions if needed, or automating this process. Testing consistency of formulas from Table 1 and formulas describing current system state enables also forecasting of situations.

4   Implementation

The intension of developers to create cross-platform easily maintainable and reusable applications resulted, in particular, in a wide use of Web technologies. Web applications which at the beginning were meant to deliver static HTML code (hypertext), nowadays possess a lot of powerful tools (HTML5, ECMAScript 6, various libraries and frameworks for client and server tasks, virtualization, event management, resource interconnection) and are used in a wide range of complex applied tasks like document management (Google Docs service), creation of 3D maps and models (WebGL, BabylonJS), videoconferencing [17]. Web interfaces are used to manage supercomputing systems, cloud services. Data presentation standards (XML, JSON) are developed on the basis of Web technologies.

Other features of Web applications are low hardware and software requirements, centralized updates, wide possibilities in distributed system development, stable network standards. This makes Web technologies a perspective, but insufficient tool for smart space development, because there is a need for dynamic models and means of implementing the service behavior scenarios, as well as for interfaces, protocols and architectures of software for service integration.

To build the smart space as a distributed multicomponent application on the basis of the Web one can use the "Model-View-Controller" design pattern, where "model" stands for the inner state of the smart space, "view" consists of the components connecting the smart space to the outer world (input and output devices, multimodal interfaces, moving items), and "controller" coordinates "model" and "view". The "model" includes the application logic responsible for achieving the goal of the smart

Table 2. The application structure

Components	Layers
Model	**Application layer** is responsible to achieve the goals of the service as a whole
	Service layer is responsible to achieve the goals of particular subservices
Controller	**Signal layer** transfers signals (events) between model and view, and between model components
View	**Interface layer** provides the communication interface that is used by the physical layer components to connect to other items of smart space
	Physical layer contains initial components to include into the smart space

space in whole and of its subservices (for example, identification, informing, etc.). The "view" consists of original software and hardware components and possesses the interface layer that enables them to connect to the "controller" components.

Thus, the application can be represented as the following set of layers (Table 2). The layers communicate to each other raising events and publishing JSON structures that contain event information.

The services interact via unified protocols over WebSocket. In comparison to the REST (Representational State Transfer), web sockets allow holding the connection active for a long time, so connecting and disconnecting to transfer each item of data are not required. This is critical when the secure connection (TLS, HTTPS protocols) is used because the key generation lasts longer than an average operation of JSON message transferring, and using REST approach would cause significant loss of time

Table 3. The MINOS services

Service	Description
System services	
Data access	Provides access to the data that can be shared between services (for example, user accounts) and secures the access legitimacy
Repository	Provides information concerning the services registered in the system, state of the services and their interfaces
Administrator interface	Allows to manage the system using a web interface
Account management	Allows to create, request, modify and delete accounts, profiles, roles and privileges of registered users
Blackboard	Enables publishing information about service events that can be handled by other services
Scenario server	Analyzes the system state using scenario models and generates events according to the detected or forecasted situations
Applied services	
Portable service hub	Allows users to get access to the corporate services using portable devices (smartphones, tablets)
Corporate TV	Broadcasts media (video, audio, text messages) to the client devices (stationary screens, gadgets)
Corporate website	Enables integrating service applets into the organization website
Communication	Allows communicating with the smart space using multimodal technologies (voice, gestures, etc.)
Information and navigation	Provides to the client devices information about their physical location in the corporate area, about the location of corporate subdivisions (departments, laboratories), functional objects (library, museum) and workplaces, and builds routes to them
Videoconferencing	Establishes videoconference connection within the smart space
Identification	Allows to identify users in different ways: pattern recognition by means of surveillance cameras; voice recognition; identification of a user by electronic key/pass; identification on entering the operating system, local area network or the corporate website

and increase the service response time. Furthermore, web sockets make it possible to check the connection status permanently and to determine instantly the components that have disconnected from the system.

All services are divided into two categories: applied services performing tasks of the smart space, and system services responsible for applied services functioning and interconnection. The list of services and their tasks is presented in the Table 3.

Building an integrated smart space implies that the activities of all applied services are closely interconnected. At the same time, the service oriented architecture requires that the components are loosely bound. Furthermore, they are relatively independent and are not obliged to know about each other and take the presence of another services into consideration, otherwise the system architecture would lose its flexibility. In this case, it is advisable to use a blackboard service where the information on system events to be processed by other services can be published as JSON messages:

```
{
    id : <event ID>,
    class : <event class(es)>,
    source : {
        service : <service class>,
        node : <node ID>,
        signature : <digital signature of the service>
    },
    target : <service class>,
    timestamp : <time and date of the event creation>,
    expires : <time limitation>,
    status : <status: caught, waiting, expired>,
    contents : <set of properties>,
}
```

For this purpose, a request to the blackboard service is made. The request contains the JSON structure containing the message parameters. The return value is an object containing the error code and the generated message identifier that allows tracing its status.

5 Conclusion

The proposed approach allows coordinating various corporate service activities in order to achieve the common goals of the system, such as information exchange, security monitoring, etc. The mathematical model used in the approach is based on the temporal logic of actions over the objects and relationships of the corporate information system data model.

The system is in the phase of active development. Necessary infrastructure is being deployed in SPIIRAS, client and server software for the corporate television and videoconferencing services has been implemented. MINOS meets the criteria of extensibility and portability and can be used as a smart space component in organizations.

Further research implies analysis of existing TLA verification methods and determination of the optimal ways of situation detecting and forecasting.

The further service development tasks include extending the possibilities of user identification using multiple approaches based on biometry [18], RFID, gadgets, etc. This enables improving interactivity and personalization of the corporate television system by integrating it with the system of biometric and RFID user identification [19]. It is supposed to implement the management of information output via gestures and speech [20–22]. Another direction of development is to provide a navigation service, which allows the visitor to determine his/her location in the organization by scanning the QR code labels. The web application allows the visitor to navigate in organization using laptop computer and get the necessary background information.

A significant field of research is related to finding the compromise between HCI efficiency and user profile security, privacy and trust. Depending on the aims of the enterprise, this task can be solved by providing the most efficient HCI capabilities under the given security constraints or by reaching the maximum security level possible in particular HCI model.

Acknowledgements. The presented work was supported by the Russian Science Foundation (grant No. 16-19-00044).

References

1. Kopacek, T.: Development trends in robotics. In: International Federation of Automatic Control Conference Paper, pp. 36–41 (2016)
2. Toro, C., Barandiaran, I., Posada, J.: A perspective on knowledge based and intelligent systems implementation in industrie 4.0. In: 19th International Conference on Knowledge Based and Intelligent Information and Engineering Systems, Procedia Computer Science, vol. 60, pp. 362–370 (2015)
3. Lillo, I., Niebles, J.C., Soto, A.: Sparse composition of body poses and atomic actions for human activity recognition in RGB-D videos. Image Vis. Comput. **59**, 63–75 (2017)
4. Singh, D., Mohan, C.K.: Graph formulation of video activities for abnormal activity recognition. Pattern Recogn. **65**, 265–272 (2017)
5. Batchuluun, G., Kim, J.H., Hong, H.G.: Fuzzy system based human behavior recognition by combining behavior prediction and recognition. Expert Syst. Appl. **81**, 108–133 (2017)
6. Glazkov, S.V., Ronzhin, A.L.: Context-dependent methods of automatic generation of multimodal user web-interfaces. SPIIRAS Proc. **21**(2), 170–183 (2012)
7. Karpov, A., Ronzhin, A.: A universal assistive technology with multimodal input and multimedia output interfaces. In: Stephanidis, C., Antona, M. (eds.) UAHCI 2014. LNCS, vol. 8513, pp. 369–378. Springer, Cham (2014). doi:10.1007/978-3-319-07437-5_35
8. Osipov, V.Y.: Neyrosetevoe prognozirovanie sobytiy dlya intellektual'nyh robotov. Mehatronika, avtomatizatsiya upravlenie **12**, 836–840 (2015). (in Russian)
9. Encyclopedia on automatic control systems for technological processes, Chapter 1.3. http://www.bookasutp.ru/
10. Di Flora, C., Gurp, J.C., Prehofer, C.: Towards effective smart space application development: impediments and research challenges. In: Proceedings of International Conference on Pervasive Computing: Common Models and Patterns for Pervasive Computing Workshop, pp. 1–2 (2007)

11. Langeheinrich, M.: Privacy by design – principles of privacy aware ubiquitous system. In: Proceedings of UBICOMP Conference, pp. 273–291 (2011)
12. Nixon, P.A., Wagealla, W., English, C., Terzis, S.: Chapter 11. Security, privacy and trust issues in smart environments. In: Smart Environments: Technologies, Protocols, and Applications. Wiley, New York (2005)
13. Levonevskiy, D.K., Vatamaniuk, I.V., Saveliev, A.I.: Multimodal information and navigation cloud system ("MINOS") for the corporate cyber-physical smart space. Programmnaya Inzheneriya **3**, 120–128 (2017). (In Russian)
14. Demri, S., Gastin, P.: Specification and verification using temporal logics. In: Modern applications of automata theory. IIsc Research Monographs, vol. 2. World Scientific, Singapore (2009)
15. Fainekos, G.E., Girard, A., Pappas, G.J.: Temporal logic verification using simulation. In: Asarin, E., Bouyer, P. (eds.) FORMATS 2006. LNCS, vol. 4202, pp. 171–186. Springer, Heidelberg (2006). doi:10.1007/11867340_13
16. Liu, C., Orgun, M.A.: Verification of reactive systems using temporal logic with clocks. Theoret. Comput. Sci. **220**(2), 277–408 (1999)
17. Saveliev, A.I., Ronzhin, A.L.: Algorithms and software tools for distribution of multimedia data streams in client server videoconferencing applications. Pattern Recognition and Image Analysis **25**(3), 517–525 (2015)
18. Kanev, K. et al.: A human computer interactions framework for biometric user identification. In: JJAP Conference Proceedings. The Japan Society of Applied Physics, vol. 4 (2015)
19. Ivanov D., Sokolov B., Dashevsky V. RFID-based Adaptive feedbacks between supply chain scheduling and execution control. In: Proceedings of 18th IFAC World Congress, Milano (Italy), 28 August–2 September 2011 [CD]. International Federation of Automatic Control, Prague, pp. 435–440 (2011)
20. Ronzhin, A.L., Karpov, A.A., Lobanov, B.M., Tsirulnik, L.I., Jokisch, O.: Phonetic-morphological mapping of speech corpora for recognition and synthesis of Russian speech. Informacionno-upravljajushhie sistemy **25**(6), 24–34 (2006). (in Russian)
21. Kashevnik, A.M., Ponomarev, A.V., Savosin, S.V.: Hybrid systems control based on smart space technology. SPIIRAS Proc. **35**(4), 212–226 (2014)
22. Vatamaniuk, I.V., Ronzhin, A.L.: Application of digital images blur estimation methods for audiovisual monitoring. Informacionno-upravljajushhie sistemy **4**, 16–23 (2014). (in Russian)

Improving Dependability of Reconfigurable Robotic Control System

Eduard Melnik[1], Iakov Korovin[2], and Anna Klimenko[2(✉)]

[1] Southern Scientific Center of the Russian Academy of Sciences (SSC RAS),
Rostov-on-Don, Russia
[2] Scientific Research Institute of Multiprocessor Computing Systems,
Taganrog, Russia
anna_klimenko@mail.ru

Abstract. The robotic information and control systems based on performance redundancy and decentralized dispatching concepts are fault-tolerant and have a high potential in the dependability aspect. The paper deals with a dependability improvement through the configuration forming. As the load balancing improves the reliability function, the quality criteria of system configurations relates to load balancing. Configuration forming is a multicriteria and multi-constraint problem. The proposed approach is to replace monitoring and control tasks relocation criterion by delegating of task context distribution to the software components of the system. The paper contains a new simplified model of the configuration forming problem, the dependability improvement approach and simulation results being discussed briefly.

Keywords: Robotics · Mechatronic objects · Information and control system · Reconfiguration · Configuration forming · Reliability · Dependability

1 Introduction

Dependability of the robotic information and control system (ICS) is an important contemporary issue, because of wide usage of robotics and complex mechatronic objects nowadays. Robotics is a part of oil and gas production industry, chemical industries, power plants, spacecraft and aircraft, and, its dependability level is crucial very frequently. The failures of such complexes may impact the environment or lead to casualties, so a huge number of efforts has been made to improve the dependability level since the 1940s.

According to the definition, dependability is the ability to deliver service that can justifiably be trusted [1–3]. The service delivered by a system is its behavior as it is perceived by its user. The function of a system is what the system is intended to do, and is described by the functional specification. The service is correct when the system function is implemented. A system failure is an event which occurs when the delivered service deviates from the correct one.

Reliability is one of the attributes of dependability and relates to the service continuity. Practically, the measure of reliability is a reliability function – the probability that an object will be functioning beyond the specified time [4]. So, the reliability function improving is a way to achieve the dependability of acceptable level.

A. Ronzhin et al. (Eds.): ICR 2017, LNAI 10459, pp. 144–152, 2017.
DOI: 10.1007/978-3-319-66471-2_16

Besides this, the fault tolerance must be taken into account as a means of obtaining dependability. In practice, fault-tolerance is implemented by the system redundancy: first studies relate to the works of von Neumann, Moore and Shannon [5, 6], and their successors, who developed theories of using redundancy to build reliable logic structures from less reliable components. Nowadays the structural redundancy is used almost everywhere, although it has some disadvantages, which are: system weight, cost, and resource utilization issues in general [7].

Within this paper the reconfigurable ICSs with performance redundancy and decentralized dispatching (for more detailed information see [8–10]) are under consideration. The current research contains the novel approach to dependability improvement through the particular way of configuration forming. As it will be shown below, the proposed method is based on an assumption that with reducing the number of objective functions the solution quality is improved in terms of other objective functions. As the objective function presence is mandatory for the configuration forming problem, we can deliver its implementation by the third-party facilities, which also will be described precisely.

The current paper also contains an improved formal model of the configuration forming problem, which is clarified in comparison with [11, 12], a dependability improvement method description, some simulation results and discussion.

2 Reconfigurable ICS with Performance Redundancy

The main difference between structural and performance reservation approaches is that structural redundancy proposes the presence of reserve elements within the system, and performance redundancy operates with the elements with additional performance. The principles of performance redundancy are described more precisely in [10, 12]. The concept of the performance redundancy is the cause of the reconfiguration procedure design: the system deals with the redistribution of computational tasks from the faulted computational node to the operational ones. While ICS operates in the circumstances of real time, the term "configuration" becomes one of the key terms of the system.

In the scope of this paper the configuration is the way to distribute the monitoring and control tasks (MCTs) among the computational units (CU) of ICS. Such resource allocation must be implemented according to the data exchange constraints between MCTs and other constraints which are given by hardware and software system implementation.

In case of CU failure its MCTs must be launched on other CUs, and, besides, those MCTs can be allocated on more than one CU.

Such reconfiguration scheme relates to the system reliability function: the one's value depends on CU temperature, which grows with loading increasing:

$$P_{CU} = P_{CU0}^{2^{k_d \cdot D/10}},$$ (1)

where P_{CU} – reliability function value of loaded CU, P_{CU0} – reliability function value of CU without loading, k_d – temperature dependency on loading ratio, D – CU loading.

So, the spreading of MCTs between CUs with load balancing affects the system reliability, and load-balancing criteria should be included into the multicriteria objective function.

As was mentioned above, configurations are the key term of the chosen class of ICSs.

Decentralized monitoring and control of the ICS is implemented by multiagent system. Each agent is associated with its own CU, and each agent on the initialization stage of the ICS has an access to the list of possible system configurations, which are formed on the design stage of the system.

Configuration forming problem is discussed in detail in [11, 12], but it seems to be a little bit cluttered. The new simplified version of generalized configuration problem forming model will be presented below, and then the approach of dependability improvement will be presented.

3 Configuration Forming Problem

There are N MCTs with computational complexities g_i, M CUs with equal performance m_j, $U = \{u_{ij}\}$ – the percentage of j CU performance allocated for the i MCT, T – planned completion time for the N MCTs, $F = \{f_k\}$, $k \in \{1, \ldots M\}$, – the set of simultaneously failed CUs.

Through the resource allocation every MCT links to the CU, and it can be described by the following tuple:

$a_i = <j, u_{ij}, t_i>$, where j – the CU identificator, u_{ij} – the allocated resource ratio, t_i – the time of MCT i accomplishment.

So, the set $A = \{a_i\}$ determines the configuration of ICS before failure, the set $A' = \{ a'_i \}$ determines the configuration of ICS after the reconfiguration. In fact, A' is the solution of configuration forming problem, and a'_i – the tuples which describes the new MCT assignments.

The objective functions are as follows.

Firstly, the number of MCTs relocated from the operational nodes must be minimized. In other words, if there is a solution where the MCT's new assignment propose the relocation of tasks from the operational nodes, we should choose the solution, where the number of such relocations is as small as possible. This objective function can be described with the expressions given below.

Let's determine the subtraction operator for sets A and A' so that:

$$a_i - a'_i = \begin{cases} 0, & \text{if } j \text{ is equal to } j'; \\ 1, & \text{otherwise.} \end{cases} \tag{2}$$

Then:

$$F_1 = \sum_{i=1}^{N} (a_i - a'_i) \rightarrow MIN. \tag{3}$$

The optimal location in the search space of this objective function means that only MCTs from the faulted node are relocated.

The second objective function is the minimization of the eliminated MCTs. In fact, some MCTs are critical and must be saved during the reconfiguration, and some MCTs are non-critical. But from the system survivability point of view it is extremely preferable to save as much MCTs as possible. So,

$$F_2 = |A| - |A'| \to MIN. \tag{4}$$

And, finally, the dispersion of CU loadings must be minimized:

$$F_3 = \sum_{k=1}^{K} u'_{kj} - \sum_{l=1}^{L} u'_{lq} \to MIN, \ \forall j, q, \tag{5}$$

where K is the number of MCTs assigned to the CU j, L is the number of MCTs assigned to the CU q.

The main constraint is that all MCTs must be accomplished within the planned completion time T:

$$t'_i + \frac{g_i}{u'_{ij} \cdot m_j} \leq T, \ \forall i, j. \tag{6}$$

Also the failed CUs must be taken into consideration:

$$M' = M - F, \tag{7}$$

where M', M and F are the sets of CUs.

And, lastly, the bordering conditions are: all values of the variables are positive,

$$u_{ij} < 1, u'_{ij} < 1, \forall i, j.$$

At first glance the problem is similar to the k-partition problem, which has a suitable solving method, but vector objective function makes the problem np-hard with complex and non-trivial search space. Also it must be mentioned that with the increasing of objective function number the quality of solution degrades.

As the preferable attribute of the system is the load balancing, the goal of the configuration forming is to get solutions with as good load balancing as possible. At the same time the other objective functions must be taken into consideration.

It must be mentioned that Service Oriented Architecture (SOA) concept is used in contemporary ICSs, too. Services can be relocated, hence for the SOA-based ICSs there is no need to keep the MCT relocation criteria at all.

The next section contains the approach of dependability improvement description.

4 A Dependability Improvement Approach

The reliability function is one of the dependability attributes, so, with reliability improvement we increase the dependability level. Load balancing affects onto the CU reliability, hence, the solutions of the configuration problem forming should be as good as possible in terms of load balancing.

The configuration forming problem is a three-criterion in our particular case, but, perhaps, if at least one criterion is eliminated, the quality of solutions can be improved.

But the minimization of relocated tasks from the operational nodes is expedient because of MCT context data, which, in case of reconfiguration, must be transferred to the new assigned node through communicational network, and it can take unacceptable time and resources.

The concept of the approach presented is to delegate the MCT context data distribution to the CU agents and to design system configurations without criteria of MCT relocation, while these MCTs are located on the nodes, where they can be launched.

So, when the configurations are obtained, the agents form the list of data distribution and through the regular mode of ICS prepare the actual context for the possible task relocations. As a result, we have a kind of distributed data storage, which needs some additional algorithms for its functioning.

For instance, the steps described below can be done for the context data distribution.

ICS Initialization.

- CU agent searches the list of configurations for the CU ids, where current CU MCTs can be relocated in case of other node failure. The list of distribution is formed.
- If the list of distribution is not empty, the "intention" messages are sent.
- The confirmation messages are received.
- If the "intention" message was received, the confirmation message is sent to appropriate agent.

ICS Regular Mode.

- An agent takes the MCT context data and multicasts it according to the list of distribution.
- An agent receives the context data.

ICS Reconfiguration.

- The agent loads the new configuration.
- Search the list of distribution in order to deliver actual context data to the MCTs, which has become active on the current node.
- Well-timed data delivery.

5 Simulation Results and Discussion

For the simulation a random set of 25 MCTs with computational complexity 10–40 conventional units was generated. MCTs were assigned to the 10 CUs with equal performance. The cases of failures are combinations of one random failure and two random failures simultaneously. The criterion of the solution quality is the load balancing, because of its impact onto the reliability function.

Solutions were got with the simulated annealing technique (in details see [9]). It must be mentioned that the algorithm adaptation used gives the local optimums of the problem. To evaluate the quality of solutions, the equal number of SA iterations was used for both simulations.

The simulation results are given below. On Fig. 1 the maximum CU loadings are shown, Figs. 2, 3, 4, 5 and 6 are the detailed examples of the fruitful usage of the method considered.

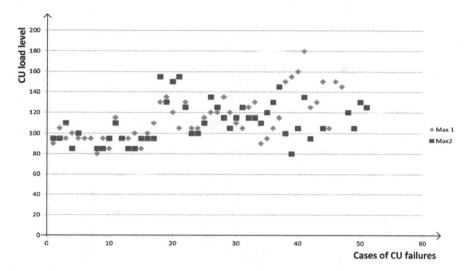

Fig. 1. Maximum CU loadings. The x-axis contains cases of CU failures, y-axis contains the loading values. Max 1 – solutions with objective functions (3, 4, 5). Max 2 – solutions with objective functions (4, 5)

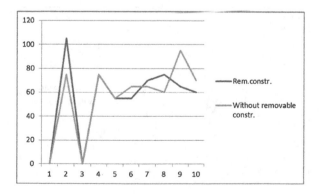

Fig. 2. CU load level when CU1 and CU3 are failed

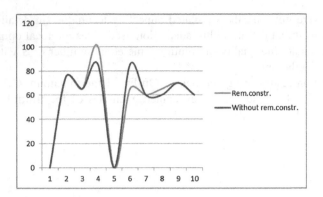

Fig. 3. CU load level when CU1 and CU5 are failed

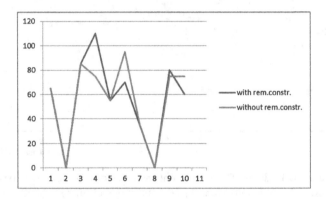

Fig. 4. CU load level when CU2 and CU8 are failed

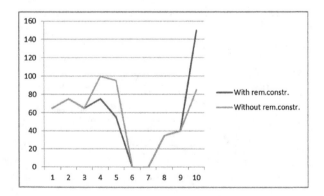

Fig. 5. CU load level when CU6 and CU7 are failed

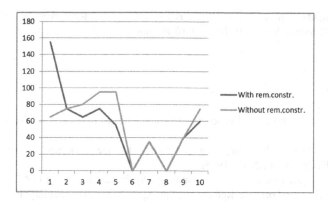

Fig. 6. CU load level when CU6 and CU8 are failed

There is a tendency on Fig. 1 of the difference of maximum loading dispersion growth with the growth of failed CU number. When the failed CU number equals 1, the maximum loadings are rather of the same magnitude. When the number of failed CUs is 2, the difference between solutions with all the criteria and without MCT relocation criterion is more obvious. It is seen that maximum loadings can be decreased, but, in some cases, the criterion removal does not produce any improvements. The probable reason of such behavior is the stochastic search particularity: with the fast, "quenching" temperature schemes the local (not global) optimums are found.

Figures 2, 3, 4, 5 and 6 contain the examples of load balancing with and without MCT relocation criterion. The cases of CU failures are shown on X-axis, and the Y-axis is the CU load level.

It is obvious that some load pikes are smoothen in the circumstances of equal SA iterations number, and we suppose that further, more precise simulations will confirm the revealed tendency.

At the same time, Figs. 1, 2, 3, 4, 5 and 6 allow to confirm, that it is expediently not to use only parallel simulated annealing search to improve the quality of solutions, but, besides this, make a search through the criteria elimination with assumption, that the criteria semantics can be delegated to the software (hardware) component of the system.

6   Conclusions

Within the scope of this paper the approach of the reconfigurable robotic control system dependability improvement was presented, described and discussed. The cornerstone of this approach is to get rid of MCT relocation criteria in multicriteria configuration forming problem and to delegate the semantic of the removed criteria to the software component of the system. Besides this, a new model of the configuration forming problem is given, simulation is done and analyzed briefly. According to the current stage of our study, it is expedient to form configurations not only with parallel search techniques (which allow to choose the best local minima), but also with the approach of "delegated" criterion. As an example, such "delegating" approach improves the solution quality up to the ratio of 1,7 (Fig. 6).

Acknowledgements. The reported study was funded by SSC RAS projects 02562014-0008, 0256-2015-0082 within the task 007-01114-16 PR and by RFBR project 17-08-01605-a.

References

1. Special session. Fundamental concepts of fault tolerance. In: Digest of FTCS-12, pp. 3–38 (1982)
2. Laprie, J.C.: Dependable computing and fault tolerance: concepts and terminology. In: Digest of FTCS-15, pp. 2–11 (1985)
3. Laprie, J.C.: Basic Concepts and Terminology. Springer, Heidelberg (1992)
4. Pham, H.: System Software Reliability. Springer Series in Reliability Engineering, 440 p. Springer, London (2006)
5. Von Neumann, J.: Probabilistic logics and the synthesis of reliable organisms from unreliable components. In: Shannon, C.E., McCarthy, J. (eds.) Annals of Math Studies, vol. 34, pp. 43–98. Princeton University Press, Princeton (1956)
6. Moore, E.F., Shannon, C.E.: Reliable circuits using less reliable relays. J. Franklin Inst. **262** (191–208), 281–297 (1956)
7. Zhang, Y., Jiang, J.: Bibliographical review on reconfigurable fault-tolerant control systems. Ann. Rev. Control **32**(2), 229–252 (2008)
8. Melnik, E., Korobkin, V., Klimenko, A.: System reconfiguration using multiagent cooperative principles. In: Abraham, A., Kovalev, S., Tarassov, V., Snášel, V. (eds.) First International Scientific Conference "Intelligent Information Technologies for Industry" (IITI 2016). Advances in Intelligent Systems and Computing, vol. 451, pp. 385–394. Springer, Heidelberg (2016). doi:10.1007/978-3-319-33816-3_38
9. Klimenko, A., Klimenko, V., Melnik, E.: The parallel simulated annealing-based reconfiguration algorithm for the real time distributed control fault-tolerance providing. In: 9 IEEE Application of Information and Communication Technologies, pp. 277–280 (2015)
10. Melnik, E.V., Korovin, I.S., Klimenko A.B.: A novel approach to fault tolerant information and control system design. In: 5-th International Conference on Informatics, Electronics Vision, University of Dhaka, Dhaka, Bangladesh (2016)
11. Korovin, I., Melnik, E., Klimenko, A.: A recovery method for the robotic decentralized control system with performance redundancy. In: Ronzhin, A., Rigoll, G., Meshcheryakov, R. (eds.) ICR 2016. LNCS, vol. 9812, pp. 9–17. Springer, Cham (2016). doi:10.1007/978-3-319-43955-6_2
12. Melnik, E.V., Klimenko, A.B.: Informational and control system configuration generation problem with load-balancing optimization. In: 10 IEEE Application of Information and Communication Technologies, pp. 492–496 (2016)

Recognition of Indoors Activity Sounds for Robot-Based Home Monitoring in Assisted Living Environments

Prasitthichai Naronglerdrit[1(✉)] and Iosif Mporas[2]

[1] Department of Computer Engineering, Faculty of Engineering at Sriracha,
Kasetsart University, Sriracha Campus, Chonburi 20230, Thailand
prasitthichai@eng.src.ku.ac.th
[2] School of Engineering and Technology, University of Hertfordshire,
Hatfield AL10 9AB, UK
i.mporas@herts.ac.uk

Abstract. In this paper we present a methodology for the recognition of indoors human activities using microphone for robotic applications on the move. In detail, a number of classification algorithms were evaluated in the task of home sound classification using real indoors conditions and different realistic setups for recordings of sounds from different locations - rooms. The evaluation results showed the ability of the methodology to be used for monitoring of home activities in real conditions with the best performing algorithm being the support vector machine classifier with accuracy equal to 94.89%.

Keywords: Sound recognition · Assisted living environments · Home robotic assistance

1 Introduction

The percentage of the elderly world population is growing and, in order to improve their quality of life and reduce the healthcare and homecare/wellbeing cost, technology-based solutions are considered. With the recent achievements of technology, ambient assisted living environments and robotic applications have been proposed for the monitoring of home activities and the support of elders or people with special needs [1–3]. Except the ageing population, indoors monitoring applications are also used for the general population in order to provide well-being by analyzing everyday activities and behaviors.

Home robot applications mainly target at monitoring of indoors human activities, the detection of hazardous situations (e.g. human body fall, or abnormal situations which have to be reported to the corresponding caregivers) and the provision of services and homecare [1–3]. The monitoring of human activities (for example cooking, eating, walking, showering, flushing etc.) is based on data acquisition from single mode or combination of multimodal sensors. The data are processed and analyzed to detect specific human activities or events of interest. Also the acquired data can be used to understand the behavior of users and their preferences when being inside their home as

© Springer International Publishing AG 2017
A. Ronzhin et al. (Eds.): ICR 2017, LNAI 10459, pp. 153–161, 2017.
DOI: 10.1007/978-3-319-66471-2_17

well as when they interact with the robot applications. Such analysis can be used to understand the affective status of the users and their daily habits.

In order to detect human activities different sensors and modalities have been employed. The collection of information can be performed either using one type of sensors or using multimodal input channels, which are afterwards typically fused on model or decision level. Also the sensors can either be static, i.e. on specific locations, or installed on moving applications, such as robots on the move.

Several sensors of different modalities have been used for human activity monitoring with one of the most common ones being the accelerometers [4–9], which measure the acceleration at an axis [10] and have proved to be quite successful in body movements detection (physical activities) and hazardous situations like falls. The main disadvantage of accelerometers is that although they can be used to monitor human body movements they are not appropriate to detect activities which are not significantly related to body moves such as watching television or singing. Except accelerometers, other modalities that have been used for human activity monitoring are wearable heart rate [11, 12] and body temperature [13, 14], vibration sensors (geophones) [15] and force sensing resistors embedded in the insole of shoe [16].

Video-based human activity monitoring is also a widely-used modality [17–20], which allows the detection and recognition of static and/or dynamic objects/schemes. In contrast to the previously mentioned sensors, video is a non-intrusive way to monitor indoors human activity. The main disadvantage of video monitoring is it's strong dependency to illumination, since during night or in dark environment video monitoring performance can become poor. Despite the use of infrared camera sensors for those cases [21], optical contact may not always be possible and thus in such cases audio-based monitoring of human activities can offer a robust solution.

The use of audio data as evidence to detect and recognize indoors human activities has been proposed in several studies found in the literature. In [22] the authors have used cepstral audio features (MFCCs) and hidden Markov models (HMMs) to recognize activities within a bathroom. In [23] the authors are using cepstral features and Gaussian mixture models (GMMs) as a classifier to discriminate sound activities. In [24] a non-Markovian ensemble voting classification methodology is proposed for the recognition of several home activity sounds. The concept of indoor sound activity simulation was introduced in [7]. In [25] the authors presented a sound-source localization methodology for the monitoring of in-home activities. The advantage of audio-based monitoring is that can operate equally well in low-illumination conditions and can – in the general case – detect an indoors event, even when barriers such as walls interject, which would be a problem in the case of video modality, and thus could be used for monitoring of assisted living environments by moving robots.

Different types of modalities have been combined, under heterogeneous sensors architectures, such as in [26] where audiovisual and medical sensors were used. In [27] video and laser scanner were deployed, while in [28] video and accelerometers were combined. In [21] physiological sensors, microphones and infrared sensors were used. Finally, in [29] the fusion of audio information with accelerometers proved to improve the automatic classification accuracy of physical activities.

In this paper we present an experimental evaluation of a sound recognition interface in home environment for sounds produced by human activities using an on-robot microphone, and investigate the appropriateness of audio-based monitoring with a single on-robot microphone in indoors human activity monitoring. In order to evaluate the sound recognition interface real indoors conditions are used. The sound recognition methodology uses time and frequency domain audio parameters and is evaluated for different classification algorithms.

The remainder of this paper is organized as follows. In Sect. 2 we present the architecture for indoors sound-based home activity recognition for robotic applications. In Sect. 3 we present the experimental setup and in Sect. 4 the evaluation results are described. Finally, in Sect. 5 we conclude this work.

2 Recognition of Indoors Human Activity Sounds

The use of microphones on robotic applications for the monitoring of indoors human activity sounds is challenging due to the fact that the microphones are not installed on fixed positions but are on the robotic system, thus can be in any room of the house. In this case we consider that the source of sound can be in one room while the sensor (i.e. the microphone) in another room. We investigate the appropriateness of one single microphone in a non-fixed position for the recognition of sounds in any room of an assisted living environment.

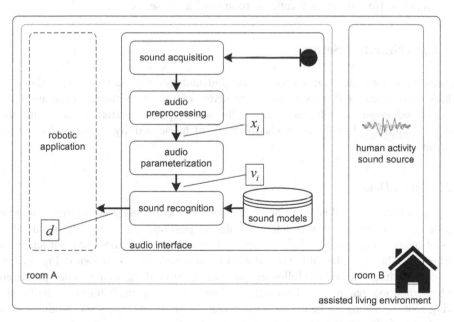

Fig. 1. Block diagram of the methodology for sound recognition of human activities in home environment. Room B is considered as the sound source location and room A as the position of the microphone sensor

The methodology for sound recognition of indoors human activities relies on a classic short time analysis and pattern recognition architecture and the evaluation focuses on the comparison of different algorithms for sound classification under different spatial setups, with respect to the position of the sound source and the position of the sensor (microphone of the assistant robot). The block diagram of the methodology is shown in Fig. 1.

As can be seen in Fig. 1, the sound source located in room B position is captured by the microphone installed on the robotic application and is processed by the audio interface module. In detail, the sound is converted to audio signal and is further pre-processed. Pre-processing consists of segmentation of the audio stream to audio frames of constant length, denoted as x_i, with $1 \leq i \leq I$. Subsequently each audio frame is processed by time-domain and frequency domain feature extraction algorithms and decomposing the signal to a sequence of feature vectors, v_i, with $v_i \in \Re^M$, and M the dimensionality of each feature vector (i.e. the total number of time-domain and frequency domain features computed). Each feature vector, v_i, is processed by a classification algorithm and using a pre-trained set of acoustic models for home sound activities recognition result d is provided to the robotic application.

After the audio-based detection of home activities or sound events of interest the robotic application will in turn act according to the application scenario, e.g. by moving towards the source of sound in order to collect visual information, by sending message to a relative or a carer/doctor etc. The present evaluation focuses on the recognition of sounds produced by home activities, under real indoors environment conditions, using a microphone from different locations (rooms) of a house.

3   Experimental Setup

In this section we describe the collection and annotation process of the audio data which were used in this evaluation, the setup of the experimental evaluation for audio-based home activity recognition, the audio parameterization stage and the machine learning algorithms for classification of home activity sounds, which were evaluated.

3.1   Audio Data

The audio data were collected in real home environment using a conventional mobile phone microphone, which was setup in different positions of the house. Specifically, sets of audio recordings with different sets of audio source – microphone positions were performed. The floor plan of the house used for the evaluation is shown in Fig. 2. The sets of recordings include the following pairs of position of the sound source produced and microphone: bathroom – bedroom, kitchen – living room, bathroom – bedroom, living room – bathroom, bedroom – bathroom. In all setups a realistic production of sounds was followed in terms of intensity and duration.

All audio data were recorded at sampling frequency equal to 16 kHz with resolution equal to 16 bit per sample. The types of home activities that were recorded as well as their duration in seconds are tabulated in Table 1. At least 10 different audio recordings for each home activity were performed.

Fig. 2. Floor plan of the house used for the recordings

Table 1. Human activities collection and their total duration

Activities	Duration (sec)
Silence	72.46
Door slamming	6.52
Flushing	39.38
Mixing with spoon	23.18
Putting on table	3.18
Showering	37.34
Footstep	24.98
Switch on/off	3.34
Tap water	35.88
Vacuuming	60.28

All audio files were manually annotated using Praat software [30] by expert audio engineer.

3.2 Audio Feature Extraction

The audio waveforms were decomposed to feature vectors using a wide range of well known and widely used time and frequency domain audio parameterization algorithms.

These are: the zero crossing rate (ZCR); the 12 first Mel frequency cepstral coefficients (MFCCs) including the 0-th coefficient; the frame energy (E); 20 linear prediction coding (LPC) coefficients; the harmonics to noise ratio (HNR); the minimum, maximum and mean sample value per frame; the 25, 50 (median) and 75% percentiles of the per frame samples; the 64-bin spectral magnitude; the energy entropy; the spectral entropy, centroid, flux and roll-off; the 12-first chroma coefficients. The total dimensionality of the feature space was $M = 123$.

3.3 Classification

For the development of the sound models of the home activities we relied on a number of widely used in the area of audio processing classification algorithms. These algorithms are: the C4.5 decision tree (C4.5), the k-nearest neighbor classifier (kNN), the support vector machines (SVM) using the sequential minimal optimization implementation with polynomial kernel (poly) and with radial basis function (rbf) kernel; the multilayer perceptron neural network with one hidden layer (NN). For the development of the home activities sound models we used the WEKA software toolkit implementations [31] and for the SVM algorithm the C and gamma parameters were selected after grid search.

4 Experimental Results

The methodology presented in Sect. 3 for audio-based recognition of indoors human activities using a non-fixed positioned microphone, as in the case of a microphone installed on a robotic device, was evaluated according to the setup described in Sect. 3. The overall human activities sound recognition accuracy for different number of classification algorithms is tabulated in Table 2. In order to avoid overlapping between the training and test data, in all experiments we followed a 10-fold cross validation protocol. The best performing results are shown in bold.

Table 2. Indoors human activity sounds recognition performance (in percentages) for different classification algorithms

Classification algorithm	Accuracy
C4.5	90.10
kNN	91.13
SVM (poly)	93.78
SVM (rbf)	**94.89**
NN	93.29

As can be seen in Table 2, the best performing algorithm is the support vector machines using the rbf kernel, followed by the SVM-poly classifier. Specifically, the SVM-rbf and the SVM-poly algorithms achieved 94.9% and 93.8%, respectively. Both discriminative algorithms, i.e. the SVM and the NN achieved high performance with the neural network classification accuracy being equal to 93.29%. The C4.5 and the kNN algorithms achieved significantly worse performance when compared to the best performing SVM sound models.

In a second step we investigated the ability of the methodology in each specific room setup. In detail, we evaluated the performance of the best performing SVM-rbf model in each of the sound source position – microphone setup, as it was described in Sect. 3. The evaluation results are shown in Table 3.

Table 3. Indoors human activity recognition (in percentages) for different room setups

Classification algorithm	Sound source	Microphone	Accuracy
Silence	–	–	90.84
Door slamming	Bathroom	Bedroom	91.10
Flush	Bathroom	Bedroom	99.24
Mixing	Kitchen	Living room	99.22
Putting	Kitchen	Living room	71.07
Showering	Bathroom	Bedroom	99.14
Step	Living room	Bathroom	87.43
Switch	Living room	Bathroom	85.03
Tap	Bathroom	Bedroom	97.99
Vacuuming	Bedroom	Bathroom	96.05

As can be seen in Table 3, the best performance is 99.24% for the flushing sound recorded from the bedroom, followed by the mixing, showering, tap and vacuuming which achieved 99.22%, 99.14%, 97.99%, 96.05%, respectively. The lowest performance of the indoor human activity sound recognition was for the sound of putting objects on the table in the kitchen when recorded from the living room and was equal to 71.07%. It is worth mentioning that in most of the evaluated human activity sounds under different source-microphone setups the recognition accuracy was more than 90%, with the exception of 'putting' and 'step' and 'switch' sounds which achieved slightly lower performance. The experimental results indicate the ability of the methodology to recognize the produced sounds in a realistic setup.

5   Conclusion

We presented a methodology for the monitoring of indoors activities using audio signal captured by a microphone from different positions of the house. The setup investigated the ability of the methodology to recognize human activity based on audio evidence when been recorded from other rooms than the position of the sound source, i.e. the human activity. The experimental results indicated the validity of the methodology for all evaluated algorithms. The best achieved performance was 94.9%, in terms of overall recognition accuracy, when using support vector machines with radial basis function kernel. The evaluated scheme does not require installation of many microphones and we deem the presented methodology can support monitoring systems in which the human activity detection is performed using a conventional microphone installed on a robot on the move, in assisted living environments. The audio-based monitoring is non-intrusive to the user and allows the monitoring of activities even during night, where typical video monitoring systems are not accurate.

References

1. Do, H.M., Sheng, W., Liu, M.: Human-assisted sound event recognition for home service robots. Robot. Biomimetics **3**(1), 1 (2016)
2. D'Arcy, T., Stanton, C., Bogdanovych, A.: Teaching a robot to hear: a real-time on-board sound classification system for a humanoid robot. In: Proceedings of Australasian Conference on Robotics and Automation (2013)
3. Do, H.M., Sheng, W., Liu, M., Zhang, S.: Context-aware sound event recognition for home service robots. In: 2016 IEEE International Conference on Automation Science and Engineering (CASE), pp. 739–744. IEEE (2016)
4. Politi, O., Mporas, I., Megalooikonomou, V.: Human motion detection in daily activity tasks using wearable sensors. In: 2014 Proceedings of the 22nd European Signal Processing Conference (EUSIPCO), pp. 2315–2319. IEEE (2014)
5. Naranjo-Hernández, D., Roa, L.M., Reina-Tosina, J., Estudillo-Valderrama, M.A.: Som: a smart sensor for human activity monitoring and assisted healthy ageing. IEEE Trans. Biomed. Eng. **59**, 3177–3184 (2012)
6. Politi, O., Mporas, I., Megalooikonomou, V.: Comparative evaluation of feature extraction methods for human motion detection. In: Iliadis, L., Maglogiannis, I., Papadopoulos, H., Sioutas, S., Makris, C. (eds.) AIAI, pp. 146–154. Springer, Heidelberg (2014). doi:10.1007/978-3-662-44722-2_16
7. Ince, N.F., Min, C.-H., Tewfik, A.H.: Integration of wearable wireless sensors and non-intrusive wireless in-home monitoring system to collect and label the data from activities of daily living. In: 2006 3rd IEEE/EMBS International Summer School on Medical Devices and Biosensors, pp. 28–31. IEEE (2006)
8. Uslu, G., Altun, Ö., Baydere, S.: A Bayesian approach for indoor human activity monitoring. In: 2011 11th International Conference on Hybrid Intelligent Systems (HIS), pp. 324–327. IEEE (2011)
9. Thiruvengada, H., Srinivasan, S., Gacic, A.: Design and implementation of an automated human activity monitoring application for wearable devices. In: IEEE International Conference on Systems, Man and Cybernetics, SMC 2008, pp. 2252–2258. IEEE (2008)
10. Mukhopadhyay, S.C.: Wearable sensors for human activity monitoring: a review. IEEE Sens. J. **15**, 1321–1330 (2015)
11. Yan, L., Bae, J., Lee, S., Roh, T., Song, K., Yoo, H.-J.: A 3.9 Mw 25-electrode reconfigured sensor for wearable cardiac monitoring system. IEEE J. Solid-State Circuits **46**, 353–364 (2011)
12. Ravanshad, N., Rezaee-Dehsorkh, H., Lotfi, R., Lian, Y.: A level-crossing based QRS-detection algorithm for wearable ECG sensors. IEEE J. Biomed. Health Inform. **18**, 183–192 (2014)
13. Parkka, J., Ermes, M., Korpipaa, P., Mantyjarvi, J., Peltola, J., Korhonen, I.: Activity classification using realistic data from wearable sensors. IEEE Trans. Inf. Technol. Biomed. **10**, 119–128 (2006)
14. Leonov, V.: Thermoelectric energy harvesting of human body heat for wearable sensors. IEEE Sens. J. **13**, 2284–2291 (2013)

15. Gabriel, I.V., Anghelescu, P.: Vibration monitoring system for human activity detection. In: 2015 7th International Conference on Electronics, Computers and Artificial Intelligence (ECAI), pp. AE-41–AE-44. IEEE (2015)
16. Shad, A., Rodriguez-Villegas, E.: Proof of concept of a shoe based human activity monitor. In: 2012 Annual International Conference of the IEEE Engineering in Medicine and Biology Society (EMBC), pp. 6398–6401. IEEE (2012)
17. Zhou, Z., Dai, W., Eggert, J., Giger, J.T., Keller, J., Rantz, M., He, Z.: A real-time system for in-home activity monitoring of elders. In: Annual International Conference of the IEEE Engineering in Medicine and Biology Society, EMBC 2009, pp. 6115–6118 (2009)
18. Amiri, S.M., Pourazad, M.T., Nasiopoulos, P., Leung, V.C.: Non-intrusive human activity monitoring in a smart home environment. In: 2013 IEEE 15th International Conference on e-Health Networking, Applications & Services (Healthcom), pp. 606–610. IEEE (2013)
19. Zouba, N., Bremond, F., Thonnat, M.: An activity monitoring system for real elderly at home: validation study. In: 2010 Seventh IEEE International Conference on Advanced Video and Signal Based Surveillance (AVSS), pp. 278–285. IEEE (2010)
20. Zhou, Z., Chen, X., Chung, Y.-C., He, Z., Han, T.X., Keller, J.M.: Activity analysis, summarization, and visualization for indoor human activity monitoring. IEEE Trans. Circuits Syst. Video Technol. **18**, 1489–1498 (2008)
21. Medjahed, H., Istrate, D., Boudy, J., Dorizzi, B.: Human activities of daily living recognition using fuzzy logic for elderly home monitoring. In: IEEE International Conference on Fuzzy Systems, FUZZ-IEEE 2009, pp. 2001–2006. IEEE (2009)
22. Chen, J., Kam, A.H., Zhang, J., Liu, N., Shue, L.: Bathroom activity monitoring based on sound. In: Gellersen, H.-W., Want, R., Schmidt, A. (eds.) Pervasive 2005. LNCS, vol. 3468, pp. 47–61. Springer, Heidelberg (2005). doi:10.1007/11428572_4
23. Vuegen, L., Van Den Broeck, B., Karsmakers, P., Vanrumste, B.: Automatic monitoring of activities of daily living based on real-life acoustic sensor data: a preliminary study. In: Fourth Workshop on Speech and Language Processing for Assistive Technologies (SLPAT) Proceedings, pp. 113–118. Association for Computational Linguistics (ACL) (2013)
24. Stork, J.A., Spinello, L., Silva, J., Arras, K.O.: Audio-based human activity recognition using non-Markovian ensemble voting. In: RO-MAN, pp. 509–514. IEEE (2012)
25. Bian, X., Abowd, G.D., Rehg, J.M.: Using Sound Source Localization to Monitor and Infer Activities in the Home. Georgia Institute of Technology, Atlanta (2004)
26. Marsh, A., Biniaris, C., Velentzas, R., Leguay, J., Ravera, B., Lopez-Ramos, M., Robert, E.: A multi-modal health and activity monitoring framework for elderly people at home. In: Yogesan, K., Bos, L., Brett, P., Gibbons, M.C. (eds.) Handbook of Digital Homecare, pp. 287–298. Springer, Heidelberg (2009). doi:10.1007/978-3-642-01387-4_14
27. Motamed, C., Lherbier, R., Hamad, D.: A multi-sensor validation approach for human activity monitoring. In: 2005 8th International Conference on Information Fusion, 8 pp. IEEE (2005)
28. Tao, L., Burghardt, T., Hannuna, S., Camplani, M., Paiement, A., Damen, D., Mirmehdi, M., Craddock, I.: A comparative home activity monitoring study using visual and inertial sensors. In: 2015 17th International Conference on E-health Networking, Application & Services (HealthCom), pp. 644–647. IEEE (2015)
29. Ketabdar, H., Qureshi, J., Hui, P.: Motion and audio analysis in mobile devices for remote monitoring of physical activities and user authentication. J. Location Based Serv. **5**, 182–200 (2011)
30. Boersma, P., Weenink, D.: Praat: Doing Phonetics by Computer (Version 5.3.51) [Computer Program] (2009). Accessed 1 May 2009
31. Witten, I.H., Frank, E., Hall, M.A., Pal, C.J.: Data Mining: Practical Machine Learning Tools and Techniques. Morgan Kaufmann, Burlington (2016)

Stationary Device for Drone Detection in Urban Areas

Petr Neduchal, Filip Berka, and Miloš Železný[(✉)]

Faculty of Applied Sciences, New Technologies for the Information Society,
University of West Bohemia, Univerzitní 8, 306 14 Pilsen, Czech Republic
{neduchal,zelezny}@ntis.zcu.cz, berkaf@kky.zcu.cz

Abstract. Drones became popular and useful in the last years. There are a lot of companies engaged in the unmanned aviation as well as ordinary people who want fly for fun. The problem arises when a drone is used against the law. Thus, the question of people protection against flying drones has to be addressed. This paper is focused on the creation of the stationary drone detection device suitable for using in urban areas. The main contribution of the paper is a detection approach composed of three parts. The motion detection, object description and classification. Moreover, the Robot Operating System is used in the proposed system in order to create an easily modifiable system. The performance of the proposed approach is tested in a couple of experiments. The system is able to distinguish a drone from e.g. a car or a walking person and it is able to work in real time.

Keywords: Image processing · Drone detection · Motion detection · Object classification · Camera · Area monitoring

1 Introduction

Drones became popular in the last years. They can be useful in many ways. It can be a leisure activity for vacation or it can be used in wide range of industry and agriculture for example for mapping, exploration or search and rescue tasks. Unfortunately, drones can be used against the law. For example, it can be used to spy someone or flying over urban areas without permission and in fact threaten people walking under the drone. Thus, the question of people protection has to be addressed. The solution has two parts in general. The first one is a device capable of detecting a flying object and then decides whether it is a drone or a bird for example. The second part is a device capable to safely eliminate drone. This paper is focused on the first part.

In the case of our research, the problem is defined as a creation of stationary drone detection device designed for using in the urban areas. The device should be able to detect drone in 360° around itself. It can be done by several types of sensors such as cameras, audio sensors or radars. Our research is currently based on multiple cameras.

© Springer International Publishing AG 2017
A. Ronzhin et al. (Eds.): ICR 2017, LNAI 10459, pp. 162–169, 2017.
DOI: 10.1007/978-3-319-66471-2_18

The task of camera based drone detection in the urban areas contains several challenges. For example, the area is usually large so it can be difficult to detect approaching drone in time – before it arrives too close to protected area. Another challenge arises from the diversity of the scene background. It is simple to detect drone against a clear sky. More difficult is to detect drone against trees or another heterogeneous background. Examples of drones flying in front of different backgrounds are shown in Fig. 1. The differences between various backgrounds are clearly visible. Drones are easy to recognise in front of a sky background. On the other hand, backgrounds such as a chimney, a building or a tree significantly reduce the chance to detect flying drone.

Fig. 1. Examples of drones in front of various backgrounds

The goal of this paper is to propose a system suitable to perform drone detection and classification tasks in the urban areas. The main contribution is an approach to the detection of the drone by the stationary device using multiple cameras. Moreover, the created framework will be reusable with a different number of cameras without changes in the core of created software approach. It is achieved by using Robot Operating System (ROS) [3,4] as a middleware for communication with hardware and between individual parts of proposed system.

The paper is structured as follows. In the Sect. 2, the related work associated with the problem of this paper or its particular part will be mentioned. The overview of the proposed system will be described in the Sect. 3. Section 4 is focused on the results obtained during experiments. The future work and system results are finally discussed in Sect. 5.

2   Related Work

They are related work in the field of drone detection. The first category is products of companies that offer detector devices such as Dedrone or Arronia etc. Devices are usually oriented into one particular direction – their field of view is up to 180° – and they used multiple sensors such as cameras, infrared camera sensor, audio sensors. Some of these sensors can significantly increase the cost of the product.

The research that is slightly similar to our paper can be found in papers of Rozantsev et al. [5,6]. The problem that is solved in the papers are the drone detection but they assume that the camera is not stationary. Therefore, different methods of the detection of moving objects have to be applied. Authors proposed data structure called st-cubes. It is a cube composed of n slices, where n is a number of image frames. Every slice is a part of the frame with the drone in the centre. We do not need this structure because we can use motion detector without compensation of the moving camera instead.

3   System Overview

In this chapter, we firstly describe the design of the system and then go through the methods we used for drone detection. The proposed system should be capable of detecting approaching drones as far as possible and be able to handle multiple camera stream sources. An example of the function of such a device is shown in Fig. 2. In order to create the system that will be customizable by the number of attached camera, used algorithms or visualisation, we decide to use ROS. It is a useful open source software framework with a large number of hardware drivers, third-party applications and powerful system of messages containing data that can be sent between applications written for ROS. It allows us to create algorithms as individual apps. The consequence is that the created system is multiprocess. Moreover, such a system is easy to extend because a new application can be attached to some message without changes in other applications.

An important and rather unfortunate thing one has to bear in mind is that the maximum distance where the system is able to detect a drone goes hand in hand with the resolution of the cameras. With higher resolution cameras it should be able to detect a drone at longer distances and vice versa. On the other hand, higher resolution cameras could increase the processing time. A compromise must be made then. Therefore, we assume that the drone will fly into camera view. We detect this motion by using difference image and search the area whether it is a drone moving or not.

3.1   Flying Object Detection

It is necessary to perform three steps in order to detect a flying object and classify whether it is a drone or not. The first step is motion detection which is

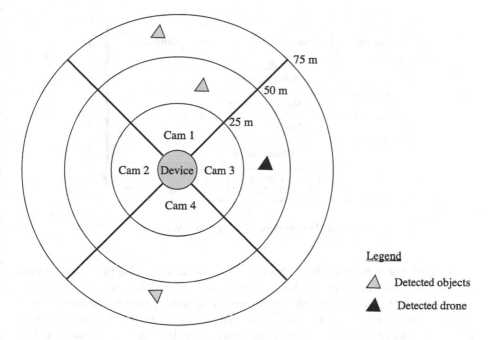

Fig. 2. Example of drone detection device function

responsible for the detection of every moving object in the image. In the second step detected image areas are described in order to classify them in the third part of the flying object detection task. In Fig. 3, there is an overview of the system design. In the first step, a frame is grabbed from all attached cameras. Then all three mentioned parts of our approach is performed. The last step is a visualisation of detected object and evokes appropriate reaction.

Motion Detection. As we mentioned earlier we use stationary cameras. Because of that, the task of motion detection [8] becomes a lot easier than with movable cameras. A difference image between two consecutive frames can then be taken into consideration.

The difference image is computed as the absolute difference of previous frame and the current one. The result is then thresholded in order to eliminate background noise which can cause a number of false positive detections. Thus, only the important changes are considered in the resulted binary image. These images are then used to select regions of interest which are then passed to the next part of the system. They are defined as bounding boxes around the detected area. We assume that in most cases the boxes containing a drone will have its width greater than its height – the aspect ratio will be greater than one. For example, for a walking person, the ratio of bounding box should be lower than one. Such boxes can be then ignored in order to improve system performance.

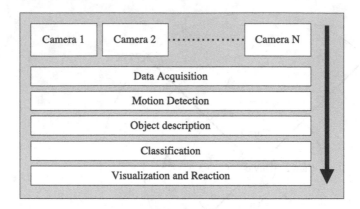

Fig. 3. Drone detection system overview

In Fig. 4, there are two examples of difference images. The first two column represent previous and current frame respectively. The third column represents the absolute difference between the frames. The fourth column represents the difference image after thresholding. The second row shows what happens when the moving object is farther away from the camera. It is clearly visible that the region of interest becomes smaller.

Fig. 4. Examples of difference images

Object Description. As it is stated before we want to search the detected regions of interest whether it was a drone moving there or not. This leads to the problem of classification because the motion detection part can detect other things like birds or planes as well. Such a detections are false positives and have to be discarded. For a successful classification, a good description of the object must be provided. We chose a similar approach as in [5]. We recorded our own videos with stationary cameras and annotated some of them by hand. Examples of the data can be seen in Fig. 5. The first row shows positive data with a drone in the centre. We tried to capture the drone in front of many different backgrounds and in many scales. The second row shows negative data. These contain other things such as cars, humans, birds, general background, and etc.

Fig. 5. Examples of training images

We assume that drones have a somewhat unique shape so we use Histogram of Oriented Gradients [1,2] as an object descriptor. The size of the training images varies based on the distance of the drone. That is a problem because feature vectors are supposed to have the same dimension. Because of that, we resized all of them to the size of 120×48 as can be seen in Fig. 5. The HoG is then computed for each image with 9 orientations in 8×8 cells. The block size is set to 3×3. Thus, 4212–dimensional feature vector is obtained.

Classification. The next step of the system is classification. We decided to use Support Vector Machine (SVM) [7,9] as a classifier because it is well-known and easy to use approach and it is frequently used in combination with HoG as features [1,2]. We take each of the bounding boxes we found with the motion detector and resize it to match the size of the training data. Then one feature vector is computed and classified by the SVM.

4   Results

In this section, we show results we achieved with the system. Figure 6 shows examples of the system in action. It performs three steps described in Sect. 3.1. At first, the motion is detected and bounding boxes around detected areas are constructed. Then these areas are resized and HoG feature vectors are computed. Thus, the objects are described. The third step is classification by SVM. If there is a drone in the bounding box the system draws a green rectangle. Otherwise, the system draws the rectangle of red colour. It is shown in Fig. 6. The walking person is labelled by the red rectangle and the drone by the green one.

The processing time of the method is dependent on the number of regions that are returned from the motion detection part. On average it takes around 18 ms to process one 540×960 frame of our test video. So the system is able to work in real-time when processing stream from a single camera. The speed performance will be lower with increasing of the camera resolution and the number of attached cameras.

In this particular video from which the images in Fig. 6 were taken we reached accuracy of 88.6%. This score was computed as a ratio of number of green bounding boxes containing a whole drone and the number of green bounding boxes not containing a drone. The errors were mainly caused by unnoticeable movement of the drone or by the drone being in front of something of similar colour as the drone.

Fig. 6. Examples of classified regions where motion was detected (Color figure online)

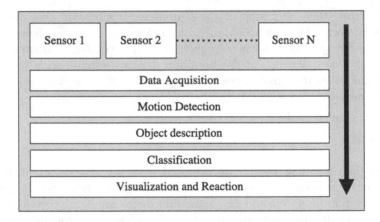

Fig. 7. The overview of the future version of our drone detection system

5   Conclusion and Future Work

We proposed a system with the ability of drone detection in urban areas. The detection is based on motion detection using difference image and classification of the moving object. The system performs fairly well if the drone is moving quickly so the motion detector captures it and it is possible to crop out the region and classify it. When the drone stops and hangs in the air no motion is detected. Therefore, no classification takes place and the system loses information about the drone in the scene. This problem will be addressed in the future.

Also, a more sophisticated approach could be chosen for the classification part of the system using some kind of convolutional neural network for the feature extraction. These machine learned features often perform better than the hand–crafted ones. Then we could perhaps use a detection algorithm based on multi–scale sliding window and classification without being dependent on the difference image.

In the future work, the proposed system will be also extended by attaching more types of sensors. Particularly we want to test audio and radar sensors. These sensors will be used in detection part of the system for improving detection rate against a heterogeneous background such as trees. It is worth mentioning that system is easy to extend because of ROS. The overview of the planned system is shown in Fig. 7 – the main change is the capability of the system to work with multiple types of sensors without significant changes in the system.

The second topic of future work in this area is the annotation of recorded data in order to create a dataset which can be used for benchmarking drone detection algorithms.

Acknowledgments. This publication was supported by the project LO1506 of the Czech Ministry of Education, Youth and Sports, by the grant of the University of West Bohemia, project No. SGS-2016-039.

References

1. Dalal, N., Triggs, B.: Histograms of oriented gradients for human detection. In: 2005 IEEE Computer Society Conference on Computer Vision and Pattern Recognition (CVPR 2005), vol. 1, pp. 886–893, June 2005
2. Dalal, N., Triggs, B.: Object detection using histograms of oriented gradients. In: Pascal VOC Workshop, ECCV (2006)
3. Garage, W.: Robot operating system (ROS) (2012)
4. Quigley, M., Conley, K., Gerkey, B., Faust, J., Foote, T., Leibs, J., Wheeler, R., Ng, A.Y.: Ros: an open-source robot operating system. In: ICRA Workshop on Open Source Software, vol. 3, p. 5, Kobe (2009)
5. Rozantsev, A., Lepetit, V., Fua, P.: Flying objects detection from a single moving camera. In: Conference on Computer Vision and Pattern Recognition (CVPR) (2015)
6. Rozantsev, A., Lepetit, V., Fua, P.: Detecting flying objects using a single moving camera. IEEE Trans. Pattern Anal. Mach. Intell. **39**, 879–892 (2017)
7. Schölkopf, B., Smola, A.J.: Learning with Kernels: Support Vector Machines, Regularization, Optimization, and Beyond. MIT Press, Cambridge (2002)
8. Sonka, M., Hlavac, V., Boyle, R.: Image Processing, Analysis, and Machine Vision, 4th edn., p. 912. Cengage Learning (2014). https://www.amazon.com/Image-Processing-Analysis-Machine-Vision/dp/1133593690/ref=pd_lpo_sbs_14_t_1/133-7853841-6586713?_encoding=UTF8&psc=1&refRID=5P23DD6973W85BB4VC26
9. Steinwart, I., Christmann, A.: Support Vector Machines. Springer Science & Business Media, Berlin (2008)

Detecting Distracted Driving with Deep Learning

Ofonime Dominic Okon and Li Meng[(✉)]

School of Engineering and Technology, University of Hertfordshire,
Hatfield, UK
l.l.meng@herts.ac.uk

Abstract. Driver distraction is the leading factor in most car crashes and near-crashes. This paper discusses the types, causes and impacts of distracted driving. A deep learning approach is then presented for the detection of such driving behaviors using images of the driver, where an enhancement has been made to a standard convolutional neural network (CNN). Experimental results on Kaggle challenge dataset have confirmed the capability of a convolutional neural network (CNN) in this complicated computer vision task and illustrated the contribution of the CNN enhancement to a better pattern recognition accuracy.

Keywords: Distraction detection · Accident prevention · Convolutional neural networks · Kaggle challenge · Triplet loss

1 Introduction

Driving is a complex task and requires a number of skills such as cognitive skills, physical fitness, coordination and, most importantly, attention and concentration of the driver on the driving [1, 2]. Despite of the complex nature of driving, it is common of drivers to get involved in activities that divert their full attention from driving, degrade their driving performance and even lead to fatal accidents. Typical examples of such activities include using a mobile phone, eating or drinking, using a navigation device, grooming, tuning the audio system, and/or talking to passengers, etc. In a report by the National Highway Traffic Safety Administration (NHTSA), it has been estimated that approximately 25 percent of car accidents were due to inattention of drivers [3] and around 50 percent of these accidents were caused by distraction of drivers [4, 5].

With the goal of reducing car accidents and improving road safety, various computer vision based approaches have been proposed. State Farm has initiated a competition called Kaggle competition, which aims to distinguish distracted driving behaviours from safe driving using images captured by a single dashboard camera. This paper presents a solution to the Kaggle challenge by using the latest development in machine learning and computer vision, i.e. deep learning and a convolutional neural network (CNN).

The paper is organized as follows. Section II provides a more in-depth description of the subject of distracted driving. Section III presents the existing computer vision based approaches to the detection of distracted driving. Section IV provides a brief subject review of deep learning and CNNs as well as a detailed description of the CNN

© Springer International Publishing AG 2017
A. Ronzhin et al. (Eds.): ICR 2017, LNAI 10459, pp. 170–179, 2017.
DOI: 10.1007/978-3-319-66471-2_19

we have adopted for the Kaggle challenge. Furthermore, section IV presents the details about the Triplet loss for the improvement in the accuracy of deep learning classification. Section V explains the Kaggle challenge, describes our experimental setup and compare the results of our two CNN models on the Kaggle images. Finally, Section VI concludes the paper and highlights some remaining challenges.

2 Distracted Driving

Distraction is a type of inattention. It has been defined by the American Automobile Association Foundation for Traffic Safety (AAAFTS) as the slow response of a driver in recognizing the information required to complete driving task safely due to some event within or outside the vehicle, which causes the shift of driver attention from the driving task [1, 4, 6]. Distraction can be categorized into four main types; visual distraction, auditory distraction, cognitive distraction and biomechanical distraction [7]. Visual distraction is the diversion of driver's visual field while looking within or outside the vehicle to observe any event, object or person [8]. Cognitive distraction is defined as diversion of thoughts from driving due to thinking about other events [9]. Auditory distraction is defined as diversion from driving due to the use of a mobile phone, communicating with other passengers or any other audio device [9]. Biomechanical distraction is diversion due to physical manipulation of objects instead of driving [10]. It is important to note that although distraction is categorized into four different types they do not occur individually but are usually linked with each other. For example, in the activity of answering an incoming call all four types of distractions can be observed: visual distraction when looking at the phone screen to interpret the phone alert and to locate the right button(s) to press; auditory distraction when hearing the alert and when being in the conversation; physical distraction when taking a hand off the wheel to press a button to receive the call; and cognitive distraction when diverting thoughts to the topic of conversation.

A research by the National Highway Traffic Safety Administration NHTSA stated thirteen different sources of distraction, which can be further categorized into technology based, no-technology based and miscellaneous sources [4]. Table 1 presents the common sources of distracted driving as identified by the NHTSA. As shown in Table 1, some technical enhancements in modern vehicles, such as the navigation system and the entrainment system, on one hand are assisting drivers in many ways but on the other hand have become sources of distraction to drivers. Furthermore, it has been predicted by Stutts et al. [11] that number of distraction-related accidents will increase with the enhancements of vehicle technologies.

Studies have been carried out to investigate the impact of distracted driving to car crashes. Stutts et al examined the Crashworthiness Data System gathered from 1995 to 1999 to identify the contribution of different distractions to accidents [11]. Glaze and Ellis focused their study on the distraction sources from within the vehicle and investigated their contributions to car accidents based on the troopers' crash record [12]. Table 2 presents a comparison of the outcomes of these two studies.

Table 1. Different sources of distraction in drivers categorized by NHSTA [4]

Type of distraction	Source of distraction
Technology based distraction	Operating radio of music devices
	Talking or listening on mobile phone
	Dialing mobile phone
	Adjusting climate controls
	Using device/object brought into vehicle
	Using device/controls integral to vehicle
Non-technology based distraction	Eating or drinking
	Outside person, object or event
	Other occupants in vehicle
	Moving object in vehicle
	Smoking related
Miscellaneous	Other distraction
	Unknown distraction

Table 2. Contribution of different distraction sources to vehicle crashes

Distraction type	Stutts et al. study [11]		Glaze and Ellis's study [12]	
	Distraction sources	% of crashes	Distraction source	% of crashes
Technology based	Adjusting radio, cassette, CD*	11.4	Adjusting radio, cassette, CD*	6.5
	Using/dialing mobile phone*	1.5	Using/dialing mobile phone*	3.9
	Adjusting vehicle/climate controls*	2.8	Adjusting vehicle/climate controls*	3.6
	–	–	Technology device*	0.3
	–	–	Pager*	0.1
	Total	**15.7**		**14.4**
Non-technology based	Smoking related*	0.9	Smoking related*	2.1
	Other occupant in vehicle*	10.9	Passenger/children distraction*	8.7
	Eating or drinking*	1.7	Eating or drinking*	4.2
	Moving object ahead**	4.3	–	–
	Person, object or event**	29.4	–	–
	–	–	Grooming*	0.4
	–	–	Other personal items*	2.9

(*continued*)

Table 2. (*continued*)

Distraction type	Stutts et al. study [11]		Glaze and Ellis's study [12]	
	Distraction sources	% of crashes	Distraction source	% of crashes
	–	–	Unrestrained pet*	0.6
	–	–	Document*	1.8
	Total	**47.2**		**20.7**
Miscellaneous	Other distraction	25.6	Other distraction inside vehicle*	26.3
	Unknown distraction	8.6	–	–
	Object brought in *	2.9	–	–
	Total	**37.1**		**26.3**

* Inside Vehicle Distraction Source
** Outside Vehicle Distraction Source

3   Previous Work

This section presents a review of the computer vision based approaches to distraction detection of drivers proposed by researchers in the literature.

Study of driver's visual behaviour has been widely carried out by researchers since 1960 [13]. Eye glance is considered a valid measure among researchers for the detection of distraction in drivers [14, 15]. In the eye glance approach, the frequency and the duration of a driver's eye glances for a secondary task are taken to produce a total measure of eyes off the road [13]. Eye glance of the driver can be measured by observing the driver's eye and head movements using a video sensor. Modern computer vision systems, for example FaceLAB [16], are able to provide real-time measurement of eye glance using head tracking and eye tracking techniques. In a study by Victor et al. [17], the validity of FaceLAB data as the measure for distraction detection has been studied and confirmed. Park and Trivedi [18] also applied SVR for the classification facial features to detect the distracted eye glance in drivers. Relevant facial features were extracted using the global motion approach and colour statistical analysis. Pohl et al. [19] developed a system based on the gaze direction and head position to monitor the distraction in drivers. Instantaneous distraction level was determined and a decision maker was used to classify the distraction level in drivers. Kircher et al. [20] also used the gaze direction as the measure for distraction detection and proposed two different algorithms. Murphy-Chutorian et al. [21] proposed a distraction detection system based on the head position of driver. Localized gradient histogram approach was used to extract the relevant features and were classified using Support Vector Regressor (SVR) to detect the distraction in drivers.

In an effort to provide efficient solution for accident prevention due to distraction, different researchers have proposed distraction warning/alerts systems in the literature. A forward warning system for distraction system was proposed by Hattori et al. [22], which used the idea of checking if the driver is looking at road based on the visual information captured by an in-vehicle camera. PERLOOK is the parameter proposed by

Jo et al. [23] as a measure to detect the distraction level in drivers in a similar way as the PERCLOS for drowsiness detection. PERLOOK is the percentage of time in which a driver's head is rotated or the driver is not looking at the road ahead. Higher values of PERLOOK means higher duration of distraction in driver. Nabo [24] used the Smart-Eye [25] software tool for the measurement of PERLOOK to detect the distraction in drivers.

Visual occlusion detection is another approach to detecting distracted driving. It assumes that safe driving does not require the driver to look at the road all the time and short intervals are allowed for performing other tasks, such as tuning the radio or adjusting climate controls. With this assumption, secondary tasks that can be performed within 2 s are classified as 'chunkable' and considered acceptable during driving [26, 27]. During the occluded time interval, driver can work with different control devices without getting distracted [28]. Validity of visual occlusion technique for the distraction detection is widely measured by researchers and considered promising approach for measurement of visual distraction in drivers [29–31].

4   Our Deep Learning Solution

4.1   Model A: The Baseline Convolutional Neural Network

AlexNet deep network [32], which was the winner of 2012 ImageNet challenge has been used as the baseline model (Model A) in this work. In ImageNet competition, AlexNet was trained on about 1.3 million real life images of 1000 different classes of objects and has achieved the test error rate of 15.3% [32]. Figure 1 shows the architecture of the AlexNet network that we have modified and used for the Kaggle challenge.

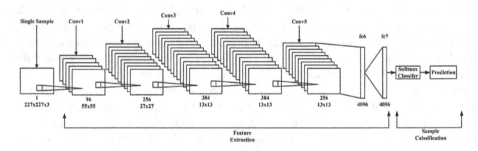

Fig. 1. Modified AlexNet deep learning architecture for Kaggle challenge

The reason behind adopting AlexNet in this work is that AlexNet (or more precisely, the architecture of AlexNet) has demonstrated its ability to learn what to 'see' in an image for the purpose of object classification. This ability means, with appropriate training, a CNN with the same architecture as AlexNet will have the ability to recognizes objects such as coke cups, phones, pets, driver's hand etc., all of which are valuable measures in classification of distracted driving.

Each input image to our AlexNet (model A) is $227 \times 227 \times 3$ as defined by the Kaggle challenge. As adopted in the ImageNet competition, the first five layers of network are convolutional layers and provide representation for local features in the images while the last layers are fully connected layers responsible for learning the key features for the given classification task. Our AlexNet extracts 4096 features at fc7 layer and creates a matrix X of the features extracted from all the training images. The dimension of feature matrix X is $m \times 4096$, where m is the number of training images in each batch. In our work, m equals 50. This extracted feature matrix is then fed into the Softmax classifier, which predicts the probabilities of the images in the input batch to the output classes. In Kaggle challenge, there are 10 classes of distracted driving. The output probability values from the Softmax classifier will be compared to the ground truth labels to calculate the following classification loss.

$$logloss = \frac{1}{N} \sum_i^N \sum_j^M y_{ij} \log(p_{ij}), \tag{1}$$

where N is the total number of images, M is the total number of classes, y_{ij} is the actual class of image and p_{ij} is the predicted class of image.

4.2 Model B: CNN Enhanced with Triplet Loss

In this work, triplet loss has been used to fine tune the model A network pre-trained with classification loss to improve the overall accuracy of the model. There are three main components in each triplet, a positive, an anchor and a negative sample as shown in Fig. 2. The aim of applying triplet loss is to minimize the distance between the anchor and the positive during the learning process and simultaneously increases distance between the anchor and the negative during the learning process to improve the classification accuracy of deep networks. Equation 2 represents the mathematical formulation of triplet loss [33].

$$\sum_i^N \max\left(0, f\left(x_i^a, x_i^p\right) - f\left(x_i^a, x_i^n\right) + \alpha\right), \tag{2}$$

where x_i^a represents the anchor feature vector, x_i^p the positive feature vector and x_i^n the negative feature vector; and α is the forced margin between the anchor-to-positive distance and the anchor-to-negative distance. $f\left(x_i^a, x_i^p\right)$ is the function which gives the distance between two feature vector. Triplet loss function from this equation tries to set

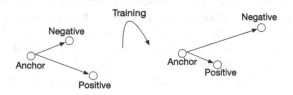

Fig. 2. Working illustration of triplet loss

apart the position samples from the negative samples by a minimum margin of α. The only condition at which the triplet loss will be greater than zero is when $f\left(x_i^a, x_i^p\right) + \alpha > f\left(x_i^a, x_i^n\right)$.

Random selection of triplets is a slow process and not much efficient for training the network. Triplets that actively contribute to the loss function and hence to improving the accuracy of the network are called hard triplets. Mining hard triplets is an essential step in efficient training of a CNN. Hard triplet selection can be done either offline or online. In offline approach triplets are generated offline for every few steps using the network checkpoint and argmin and argmax of the data are determined. While, in online approach triplets are generated by selecting the positive/negative exemplars from mini-batch [33] during live training. To fasten the convergence of our model B network with triplet loss, offline selection of hard triplets is implemented.

5   Experiments and Results

5.1   Dataset

The Kaggle competition [34] provides a dataset of 80,000 2D images of drivers for data scientists (Kagglers) to classify. Each image in the dataset is captured in vehicle, some with occurrence of distracted activities such as eating, talking on phone, texting, makeup, reaching behind, adjusting radio, or in conversation with other passengers [35]. Table 3 shows the 10 prediction classes defined by the competition.

Overall the dataset has been divided in the ratio of 90%:10% for training and testing the proposed algorithms, respectively. This means from a total of 22424 images in all the Kaggle classes, 20182 are used to train and 2242 to test the two network models.

Table 3. Prediction classes for Kaggle task and number of images in each class [34]

Class0	Class1	Class2	Class3	Class4	Class5	Class6	Class7	Class8	Class9
Safe driving	Texting-right	Talking on phone-right	Texting-left	Talking on phone-left	Operating the radio	Drinking	Reaching behind	Hair and makeup	Talking to passenger
2489	2267	2317	2346	2326	2312	2325	2002	1911	2129

5.2   Experimental Results

This section presents the results of the experiments performed to test the classification accuracy of the two proposed deep learning models as explained in Sect. 4. Overall 5000 maximum iterations were allowed to train the. Figure 3 presents the test accuracy and the test loss of both Models (A: AlexNet+Softmax and B: AlexNet+Triplet Loss) for 5000 iterations with an iteration interval of 500. It has been observed that over the number of iterations classification accuracy improved and both models converged.

Table 4 summarize the results of both algorithms after 5000 iterations. Classification accuracy of 96.8% and 98.7% has been achieved for Model A and Model B, respectively. It is important to mention here that 100% accuracy was achieved for these algorithms when applied to training dataset.

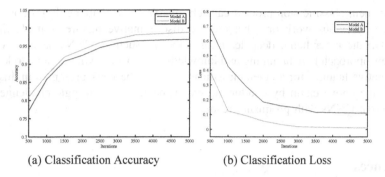

(a) Classification Accuracy   (b) Classification Loss

Fig. 3. Classification accuracy and classification loss plots of model A and model B

Table 4. Summary of experimental results for model A and model B

	Test accuracy	Test loss
Model A: AlexNet+Softmax	96.8	0.11
Model B: AlexNet+Triplet Loss	98.7	0.01

5.3   Kaggle Scores

Kaggle provided 22424 images to participants for training their algorithms and asked to submit their classification probabilities for each image in form of excel sheet. Further they tested the submitted algorithms on 79,726 un-labeled images and calculated the loss score for each participant. Kaggle evaluated each submission using a multiclass logloss function as given in Eq. 1.

Classification results from the Model A were submitted to Kaggle and were evaluated for the Kaggle score and rank. Table 5 shows the Kaggle submission results for Model A. The rank was determined at the time of submission out of approximately total 2000 submissions.

Table 5. Kaggle submission results for model A

	Kaggle score	Rank
Model A: AlexNet+Softmax	1.54860	500+

6   Conclusion and Future Works

As discussed in Sects. 2 and 3, majority of the existing approaches to the detection of distracted driving relay on information such as eye glance direction and head movement. To estimate such information, methods have been proposed for the extraction of relevant key features from the face/head region of the driver. However, the image data of the Kaggle challenge are provided for classification of different types of behaviors that involve whole body movements of the driver. To complete the Kaggle challenge, one has to first define the discriminative features from the entire body of the driver that the subsequent classification process can rely on. This is a challenging task as there is hardly any previous work on what are the discriminative features outside the face region. On the other hand, deep learning networks such as CNNs have provided a brand new approach to data mining and knowledge discovery, which is able to learn the discriminative features for a given classification task. The work presented in this paper confirms the above claim by conducting experiments on the Kaggle challenge using two different CNNs with promising results.

References

1. Beirness, D.J., Simpson, H.M., Pak, A.: The road safety monitor: driver distraction (2002)
2. Peters, G.A., Peters, B.J.: The distracted driver. J. R. Soc. Promot. Health **121**, 23–28 (2001)
3. Young, K., Regan, M., Hammer, M.: Driver distraction: a review of the literature. Distracted Driving, 379–405 (2007)
4. Stutts, J.C., Reinfurt, D.W., Staplin, L., Rodgman, E.A.: The Role of Driver Distraction in Traffic Crashes. Report prepared for AAA Foundation for Traffic Safety, Washington (2001)
5. Wang, J.-S., Knipling, R.R., Goodman, M.J.: The role of driver inattention in crashes: new statistics from the 1995 crashworthiness data system. In: 40th Annual Proceedings of the Association for the Advancement of Automotive Medicine, p. 392 (1996)
6. Treat, J.R.: A study of precrash factors involved in traffic accidents. HSRI Research Review (1980)
7. Ranney, T.A., Mazzae, E., Garrott, R., Goodman, M.J.: NHTSA driver distraction research: past, present, and future. In: Driver distraction internet forum (2000)
8. Hajime, I., Atsumi, B., Hiroshi, U., Akamatsu, M.: Visual distraction while driving: trends in research and standardization. IATSS Res. **25**, 20–28 (2001)
9. Line, D.: The Mobile Phone Report: A Report on the Effects of Using a Hand-Held and a Hands-Free Mobile Phone on Road Safety. Direct Line Insurance, Croydon (2002)
10. Haigney, D.: Mobile Phones and Driving: A Literature Review. RoSPA, Birmingham (1997)
11. Stutts, J., Feaganes, J., Rodgman, E., Hamlett, C., Meadows, T., Reinfurt, D., Gish, K., Mercadante, M., Staplin, L.: Distractions in everyday driving (2003)
12. Glaze, A.L., Ellis, J.M.: Pilot study of distracted drivers. Transportation Safety Training Center for Public Policy (2003)
13. Farber, E., Foley, J., Scott, S.: Visual attention design limits for ITS in-vehicle systems: the society of automotive engineers standard for limiting visual distraction while driving. In: Transportation Research Board Annual General Meeting, Washington DC USA, pp. 2–3 (2000)
14. Haigney, D., Westerman, S.: Mobile (cellular) phone use and driving: a critical review of research methodology. Ergonomics **44**, 132–143 (2001)
15. Curry, R., Greenberg, J., Blanco, M.: An alternate method to measure driver distraction. In: Intelligent Transportation Society of America's Twelfth Annual Meeting and Exposition (2002)
16. Seeingmachines. http://www.seeingmachines.com/. Accessed 11 Mar 2017
17. Victor, T., Blomberg, O., Zelinsky, A.: Automating the measurement of driver visual behaviours using passive stereo vision. In: Proceedings of International Conference Series Vision Vehicles (VIV9) (2001)
18. Park, S., Trivedi, M.: Driver activity analysis for intelligent vehicles: issues and development framework. In: Intelligent Vehicles Symposium Proceedings, IEEE, pp. 644–649 (2005)

19. Pohl, J., Birk, W., Westervall, L.: A driver-distraction-based lane-keeping assistance system. Proc. Inst. Mech. Eng. Part I: J. Syst. Control Eng. **221**, 541–552 (2007)
20. Kircher, K., Ahlstrom, C., Kircher, A.: Comparison of two eye-gaze based real-time driver distraction detection algorithms in a small-scale field operational test. In: Proceedings of 5th International Symposium on Human Factors in Driver Assessment, Training and Vehicle Design, pp. 16–23 (2009)
21. Murphy-Chutorian, E., Doshi, A., Trivedi, M.M.: Head pose estimation for driver assistance systems: a robust algorithm and experimental evaluation. In: Intelligent Transportation Systems Conference, pp. 709–714. IEEE (2007)
22. Hattori, A., Tokoro, S., Miyashita, M., Tanaka, I., Ohue, K., Uozumi, S.: Development of forward collision warning system using the driver behavioral information. SAE Technical paper (2006)
23. Jo, J., Lee, S.J., Jung, H.G., Park, K.R., Kim, J.: Vision-based method for detecting driver drowsiness and distraction in driver monitoring system. Opt. Eng. **50**, 127202–127224 (2011)
24. Nabo, A.: Driver Attention—Dealing with Drowsiness and Distraction. IVSS, Göteborg (2009)
25. Yunqi, L., Meiling, Y., Xiaobing, S., Xiuxia, L., Jiangfan, O.: Recognition of eye states in real time video. In: International Conference on Computer Engineering and Technology, pp. 554–559 (2009)
26. Karlsson, R., Fichtenberg, N.: How different occlusion intervals affect total shutter open time. In: Presentation at the Exploring the Occlusion Technique: Progress in Recent Research and Applications Workshop, Torino, Italy (2001)
27. Green, P., Tsimhoni, O.: Visual occlusion to assess the demands of driving and tasks: the literature. In: Exploring the Occlusion Technique: Progress in Recent Research and Applications Workshop, Torino, Italy, p. 2004 (2001)
28. Jain, J.J., Busso, C.: Assessment of driver's distraction using perceptual evaluations, self assessments and multimodal feature analysis. In: 5th Biennial Workshop on DSP for In-Vehicle Systems, Kiel, Germany (2011)
29. Baumann, M., Rösler, D., Jahn, G., Krems, J., Kluth, K., Rausch, H., Bubb, H.: Assessing driver distraction using occlusion method and peripheral detection task (2003)
30. Baumann, M., Keinath, A., Krems, J.F., Bengler, K.: Evaluation of in-vehicle HMI using occlusion techniques: experimental results and practical implications. Appl. Ergon. **35**, 197–205 (2004)
31. Wooldridge, M., Bauer, K., Green, P., Fitzpatrick, K.: Comparison of driver visual demand in test track, simulator, and on-road environments. Ann. Arbor, 1001, 48109–42150 (1999)
32. Krizhevsky, A., Sutskever, I., Hinton, G.E.: ImageNet classification with deep convolutional neural networks. In: Advances in Neural Information Processing Systems, pp. 1097–1105 (2012)
33. Tutorial: Triplet Loss Layer Design for CNN. http://www.cnblogs.com/wangxiaocvpr/p/5452367.html. Accessed 19 Mar 2017
34. Kaggle Competition: State Farm Distracted Driver Detection. https://www.kaggle.com/c/state-farm-distracted-driver-detection. Accessed 12 Apr 2017
35. Liu, D., Sun, P., Xiao, Y., Yin, Y.: Drowsiness detection based on eyelid movement. In: Second International Workshop on Education Technology and Computer Science (ETCS), pp. 49–52 (2010)

An Advanced Human-Robot Interaction Interface for Teaching Collaborative Robots New Assembly Tasks

Christos Papadopoulos[✉], Ioannis Mariolis,
Angeliki Topalidou-Kyniazopoulou, Grigorios Piperagkas,
Dimosthenis Ioannidis, and Dimitrios Tzovaras

Centre of Research and Technology – Hellas,
6th km Charilaou - Thermi, 57001 Thessaloniki, Greece
pap-x@iti.gr

Abstract. This paper presents an advanced human-robot interaction (HRI) interface that allows teaching new assembly tasks to collaborative robots. The interface provides the proper tools for a non-expert user to teach a robot a new assembly task in a short amount of time through advanced perception and simulation technologies. The assembly is demonstrated by the user through an RGBD camera and the system extracts the needed information for the assembly to be simulated and performed by the robot, while the user guides the process. The HRI interface is integrated with the ROS framework and is built as a web application that can be operated through a PC tablet. We evaluate the interface with user experience rating from test subjects that are requested to teach an assembly to the robot.

Keywords: HRI web interface · Teaching robotic assembly · Robot simulation

1 Introduction

A significant challenge facing the effort to use robots for complex assembly tasks is reducing the amount of time and resources needed to teach the robots how to perform the assembly in question. Expert roboticists can program new policies and skills within specialized domains such as manufacturing and lab experimentation, but this approach requires large amount of time and resources that are not always available [1]. Learning from demonstration has been proposed as a potential solution to this problem [2]. Using a Human-Robot Interaction (HRI) interface, the teacher, provides demonstrations of a desired task, which are then used to plan the robot actions that needs to be performed in order to successfully complete the assembly task.

This paper focuses on the specifics of the functionality of the HRI system which is used for the demonstration of the assembly from an inexperienced user and the simulation of the assembly task in a virtual environment. Although Learning from Demonstration (LfD) has been already used as a technique to teach a robot new skills [3], to the best of our knowledge it has never been used for teaching robotic assembly tasks. The corresponding HRI interface for facilitating such teaching should be simple

© Springer International Publishing AG 2017
A. Ronzhin et al. (Eds.): ICR 2017, LNAI 10459, pp. 180–190, 2017.
DOI: 10.1007/978-3-319-66471-2_20

enough so that a non-expert user can demonstrate new assembly tasks, while still enabling the user to supervise the assembly execution.

Another problem with the majority of existing HRI systems is that they require special architectures and complex interfaces to allow the user to interact with the robot [4], something that adds more difficulties to the inexperienced user. To tackle this issue, we propose the use of a simple web interface that allows the user to interact freely with the HRI system, without the constraints of a specialized architecture. The user can control complex actions of the system and supervise the process through a lightweight graphical interface on a web browser using touch controls (using a tablet PC for instance). This approach supports the creation of a user-friendly robot control interface used for demonstration and simulation of complicated assembly tasks. One of the advantages of the proposed system is that it allows the interaction of an inexperienced non-expert user with a complex robotic system for teaching assembly tasks that previously required special policies and skills from experts in this field.

The main contribution of this work can be summarized as follows:

- User-friendly and intuitive HRI interface used for teaching new assembly tasks.
- Web-based interface that is portable and integrated with ROS.
- Assembly simulation functionalities.

1.1 Related Work

Human-robot interaction is a rapidly evolving field that has applications in almost all robotic tasks including manufacturing, aviation, surgery, agriculture and education. Specialized robots under human teleoperation or teaching have proven successful in hazardous environments and medical applications, as have special tele-robots under human supervisory control for repetitive industrial tasks [5]. Research for the way humans can safely and robustly interact with robots is yet at initial stages and there is much room for improvements on this field. Especially in factory assembly lines, much work has been done on teaching robots how to perform certain tasks, such as aircraft assembly, where there have been developments of techniques for observing human subjects in performing a manipulation task [6]. Most observation from demonstration methods are performed using computer vision with different detection techniques [7], while kinesthetic learning is also a valid method for teaching a robot numerous assembly tasks [8]. On the other hand, safety in the form of collision avoidance is a continuing issue when the interaction involves autonomous robots that operate in crowded environments [9].

Recently, there has been several projects addressing the control of robots through web interfaces. Such an approach is the PR2 remote lab that is used for shared development for learning from demonstration [2]. The web client of the project features many JavaScript (JS) widgets that connect to ROS [10] middleware through a ROS JavaScript (JS) library and websockets. The user can interact with the remote robot through predefined interfaces or small JS scripts. Another similar project addressed gesture-based remote HRI using a Kinect sensor, where a custom client/server web application is combined for remotely controlling the robot through four hand gestures [11].

Recording demonstrations through a web interface has also been used previously for demonstration capturing of mobile manipulation tasks provided by non-experts in the field [12]. That approach employed also a lightweight simulation environment to reduce unnecessary computations and to improve performance through the Gazebo simulator [13]. Robot control through a web interface has also been used to allow elderly users to have access to the multi-robot services of the Robot-Era project [14].

Moreover, there have been some attempts to build web frameworks for HRI experimentation. One such attempt is the Robot Management System (RMS), a novel framework for bringing robotic experiments to the web [15]. It provides a customizable browser-based interface for robot control, integrated support for testing of multiple study conditions, and support for both simulated and physical environments. The client interacts with the ROS algorithms through the RMS web server (HTTP) and with the physical and simulated environments through JSON requests.

The majority of the existing approaches that utilize web human-robot interfaces are focused on the creation of remote laboratories for user interaction with robots, and not so much on the use of web technologies for teaching robots new assembly tasks. In our work, we combine technologies that have been used to remotely control robots with technologies that have been used to teach robots from demonstration, in order to create a high-level HRI interface for teaching robots new assembly tasks.

2   Overall Architecture

Our system is built around the concept of a robot capable of learning and executing assembly tasks demonstrated by a human. The system will learn assembly tasks, such as insertion or folding, by observing the task being performed by a human instructor. Then, the system will analyze the task and generate a corresponding assembly program whereas 3D printable fingers tailored for gripping the parts at hand will be automatically designed. Aided by the human instructor, the robot will finally learn to perform the actual assembly task, relying on sensory feedback from vision, force and tactile sensing, as well as physical human robot interaction.

All of the above work heavily relies on smooth and robust user interaction with the system using the HRI interface. The HRI interface consists of a graphical user interface with which the user interacts through a web browser, allowing deployment on portable devices, such as a PC tablet. In the back-end of the system, a web server responds to the user requests from the web browser and performs the required actions using a state machine and the Robot Operating System (ROS) to communicate with the robotic system. The communication between the browser and the webserver is performed through JavaScript and PHP requests, while the communication between the web server and ROS is performed with the ROSLIBJS library [16]. In our implementation, the web server and the ROS environment are installed on separate workstations. However, it is possible to use a single machine for both, depending on its computing capabilities.

The HRI interface is built using a server-client architecture for increased portability and centralized control. Different technologies in both the client and the server should work in unison to create an integrated environment. The client side must be intuitive and lightweight so it can be deployed on a mobile device (a PC tablet in our case). On the other hand, the server side should be able to handle the computationally demanding work of analyzing and simulating, while handling the requests made by the user through the tablet. Last but not least, the client has to be in constant communication with the server to enable the continuous interaction with the user that is required.

On the client side, we make use of the latest technologies to provide the best user experience with our interface. Some of them include HTML5, used to build the user interface, mainly with the use of buttons and frames and JavaScript (JS) that is used to provide dynamic functionality and reduce the amount of pages the client has to load. We also use Ajax and JSON along with ROSLIBJS for the client-server asynchronous communication and, Gzweb [17] that is used to provide the user tools for viewing and interacting with the 3D simulation.

On the sever side, we make use of popular technologies with robustness in mind to create a capable and fully operating server. Some of the utilized technologies include PHP, a common and easy to use programming language that is used for developing the application and calling system or ROS services, and MySQL, a common and light-weight database that is used for storing all the information needed for the HRI interface. Finally, we employ ROS with Gazebo, the environment used for implementing the simulation along with many useful libraries that they provide (Sect. 3.4) (Fig. 1).

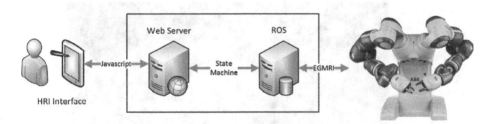

Fig. 1. System's overall architecture

3 Implementation of the Interface

The developed HRI interface consists of different modules enabling the user to teach the assembly task to the robot and supervise the learning process. The user follows a predefined sequential procedure and the graphical interface allows her to control the process. The main workflow of our system consists of three main and discrete stages (or phases) that the user should follow in order to teach a new assembly to the robot. These are the teaching phase, the design phase and the training phase.

3.1 Teaching Phase

The initial step of the teaching phase is the assembly task creation, where a name describing the assembly task can be specified by the user, the parts that the robot must assemble are selected, and the assembly type, e.g. folding assembly, has to be defined.

The models of the parts are uploaded (if they are not already in the system) and the system displays them in a 3D simulation environment using Gazebo (Sect. 3.4) (Fig. 2). The next step employs the detection of the uploaded parts in the work environment. This step is important for the demonstration process since the parts have to be identified while the user performs the assembly task. For this purpose, we use an RGBD camera streaming through a ROS node as a web video server, so the video stream can be seen by the user. The assembly parts are placed within the camera view and the system is called to identify them. A 3D representation of the object is overlaid on the image where it was recognized to provide feedback to the user for confirming the identification (Fig. 3a). After the parts have been detected, the assembly task must be demonstrated by the user in front of the RGBD camera. The video stream data are streamed through the web, using a corresponding ROS node, for the user to watch on the HRI screen. The demonstration of the assembly task can be recorded and then reviewed by the user who can choose to save it or discard it (Fig. 3b). The saved sequences of the demonstration are used to extract the placement and movement of the parts during the assembly.

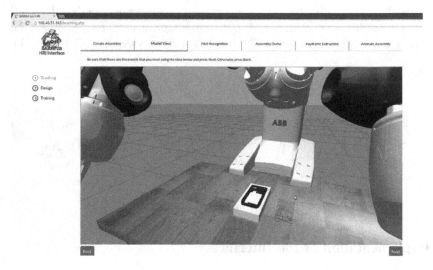

Fig. 2. The HRI interface, displaying a 3D representation of the environment with the parts

The next step of the process is performed by the Key-frame Extraction module. After the assembly task has been demonstrated and the system has captured the frames of the process, it extracts the main frames that demonstrate the movement of the parts. The user can choose to add or remove key-frames if the system's generated results do not meet the assembly's needs (Fig. 4a). Semantic information can also be added to every key-frame so the system can use it as input for the Assembly Program Generator module. Employed labels for the corresponding states are the following: Initial position, Grasping, Picking up, Moving, Aligned, Contact, Assembled, and Retract hand. The interactions between the parts and the teacher's hand are automatically identified to provide feedback to the teacher. After saving the key-frame selection, the trajectory of the parts' movement during the assembly task is created by the system.

Fig. 3. (a) Detection of the parts has been performed and the user is prompted to confirm the result, (b) The teacher's hand is detected and assembly demonstration is recorded

Last but not least, the assembly process is simulated in the Gazebo 3D simulation environment. Using the trajectory path of the parts' movement from the key-frame extraction process, the movement of the 3D objects can be animated so the user can have a preview of the actual physical assembly task (Fig. 4b). The computationally demanding processes of object detection, key-frame extraction and assembly simulation are executed on the system's server and not on the client's computer, making the graphical interface lightweight and easy to use for the non-expert user.

Fig. 4. (a) Key-frames are extracted by the system and the user provides feedback on the suggested sequence, (b) Gazebo simulation of the demonstrated assembly is generated based on the selected key-frames

3.2 Design Phase

In the design phase of the assembly, the system creates new fingers for the robot's gripper and the grasp poses for the robot. Firstly, a 3D model of the original finger of the robot's gripper is displayed and the system generates new CAD models for the fingers based on the specifications of the parts that are going to be assembled. The designing of the fingers takes place in an external system (Catia V5) and the CAD models are uploaded automatically to the HRI interface. A progress bar informs the user about the system's progress in creating the fingers and the planning of the grasps.

After the models are generated the user can inspect them through a Javascript 3D viewer and proceed to the next stage where the system presents the best grasp poses of the robot's gripper that have been generated, in the 3D environment of Gazebo. The user can view the most appropriate poses of the gripper sorted with respect to the assembly part that has to be gripped (Fig. 5a). The poses that don't seem feasible or can't be reached by the robot can be removed by the user. In this stage, the user can also request the printing of the fingers in a 3D printer and attach them to the actual robot.

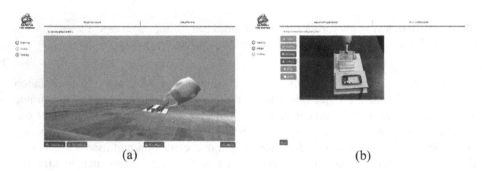

(a)   (b)

Fig. 5. (a) Display of the simulated grasp, (b) Execution of the assembly after part detection

3.3   Training Phase

In the training phase of the assembly, the appropriate Assembly Program has to be loaded for execution. After loading the program, the system informs the user that the parts have to be detected on the working table to proceed with the assembly. The detection module is similar to the one used on the teaching phase and overlays 3D representations of the objects at the position they are detected. After the detection is confirmed, the user can either ask for an assembly simulation in the Gazebo environment or proceed with the execution of the assembly. While executing the assembly, the robot motions are generated using the information that was extracted by the Key-frames extracted in the teaching phase. Due to the uncertainties that can arise about the object's position after the robot picks it up, the system requests a re-detection of the parts after the robot has picked them up. The user has a clear view of the assembly execution through a video stream that comes from the camera that is mounted on the table, so the robot movements can be inspected (Fig. 5b). The HRI interface also provides feedback about the assembly state by using messages from ROS created from the assembly program that controls the assembly process. When the assembly parts are in contact, the user has the option of stopping the execution and enabling physical-HRI to move the robot's arm to the desired position to avoid wrong positioning. Finally, when the assembly is completed the HRI interface prompts for a new object detection and re-execution of the assembly if needed.

3.4   Simulation

Simulation aids the instructor to obtain a good understanding of the assembly process as perceived by the system. Simulating various parts of the procedure in a 3D virtual environment can help avoiding errors, reducing training time, and increasing safety during operation. To aid us with the simulation of the assembly, first and foremost, we have used the Robot Operating System (ROS) that is widely used in modern robotic applications. ROS employs many tools, including algorithmic calculations and PID control simulations. Aiding the need for visualization of the simulation for various parts of our interface and providing with a 3D physics simulation environment is Gazebo. This is a simulation tool that works along with ROS and provides tools and resources for 3D simulations of robot models (after the CAD files of the robot have been obtained) and other objects such as the assembly parts. Since the developed graphical interface is going to be deployed on a PC tablet, the visualization of our simulation needs to be provided using web technologies. For this reason, we chose to use Gzweb, an alternative to Gazebo visualization client (gzclient) that runs on a web browser and provides interfaces for Gazebo.

A significant amount of effort has been put to enable simulating the robot movements using ROS and Gazebo. An invaluable tool for this purpose is ros_control, a ROS library that provides interfaces for simulating PID controllers through simple configurations and plugins. Using similar structures with the actual robot hardware, we can simulate joint movements using effort, velocity or force controllers and simulate the robot's behavior in any environment and assembly conditions. Any specific actions that had to be done for utilizing generic controllers, like the gripping mechanism or the motion generation, were created using Gazebo plugins written in C++ and python with the libraries provided by Gazebo. However, simulating a system as complex as a robot has many limitations and undefined variables like the friction values or motor transmissions, introducing discrepancies between the actual and simulated execution.

3.5   Perception

During demonstration, based on acquired image sequences, visual cues containing information about how to execute an assembly task are extracted. Recent developments allow for the use of low-cost RGBD sensors, which, include an additional channel that acquires depth images of the scene (apart from the three color channels), providing data in 2.5D. This information is regularly used by state-of-the-art methods on object detection and hand tracking in order to infer 3D pose. In our work, the 6-DoF object detection method proposed in [18] has been adopted for the detection of assembly parts that are placed inside the robot's workspace by the human teacher. During demonstration, the RGBD sequences are recorded and using this data, hand-object tracking is performed off-line. Hand tracking is based on the method proposed in [19] and the estimation of hand poses is performed using an articulated 3D model with 42 DoF.

Key-Frame Extraction. The representation of the underlying scene and spatiotemporal encoding of the teaching-by-demonstration process is achieved efficiently with key-frame extraction. Using the detection and tracking results from the perception module, the scene is segmented effectively. Pose (position and orientation) of the objects and the hands is used as input to the key-frame extraction module. For each key-frame extracted, the system exports XML files for the objects and each hand detected, containing all available semantic and technical information, such as semantics of the objects' relations, poses of objects and teacher's hands, contact points and forces, grasping states and hardware information [20].

4   Experiments and Evaluation

In this section, we present some results from the experiments we performed with 13 inexperienced users that were requested to use the interface to teach the robot an assembly task. The subjects had a brief introduction of the system's functionalities and used a tablet PC to guide the learning process for the robot. In the end, a simulation of the robot's movements was displayed. Furthermore, after completing the task, to evaluate the usability of the interface, the subjects had to answer on a five-point Likert scale the questions in Table 1.

Table 1. Questions that the subjects had to answer concerning the HRI interface

1	An inexperienced user can easily teach the robot an assembly task using this interface
2	I understood what buttons I needed to press to perform each action
3	I found the interface easy to use
4	The interface clearly guided me through the process
5	The interface presented a safe and effortless way to interact with the robot
6	The video streams and simulations gave a clear view of what was happening

The answers had the following options: strongly disagree (=1), disagree, no opinion, agree, and strongly agree (=5). The average scores and standard deviations for each question are presented in Table 2. From interviews the majority of the users evaluated positively the experience and were pretty confident with the interface if they were asked to teach a new assembly to the robot without any assistance. The average score for all the questions that the subjects had to answer was **4.31/5**.

Table 2. Median, average, and St. Deviation of question scores on a five-point Likert scale

Likert scores\question	1	2	3	4	5	6	All
Median	4	4	5	4	4	5	4
Average	4.3	4.35	4.6	3.9	4.38	4.32	4.31
St. Deviation	0.48	0.43	0.51	0.95	0.51	0.85	0.4

5   Conclusion and Future Work

In this paper, we presented the design and technical implementation of an advanced HRI interface to allow the teaching of new assembly tasks to collaborative robots. The interface is built as a web application and can be executed on tablet PCs. To this end, many advanced technologies are employed and offer a seamless experience to the inexperienced user that is requested to guide the process of Learning by Demonstration for the robot. We evaluated the usability of the system by having inexperienced users try to teach an assembly to the robot and rate the experience, for which we had positive reviews. In the future, we aim to add new features to the interface, like multiple demonstrations from different cameras for the assembly and a more generic interface to allow the construction of a broader range of assemblies.

Acknowledgment. The research leading to these results has received funding from the European Community's Framework Programme Horizon 2020 – under grant agreement No 644938 – SARAFun.

References

1. Wilcox, R., Nikolaidis, S., Shah, J.: Optimization of temporal dynamics for adaptive human-robot interaction in assembly manufacturing. Robot. Sci. Syst. **VIII**, 441–448 (2012)
2. Argall, B.D., et al.: A survey of robot learning from demonstration. Robot. Auton. Syst. **57**(5), 469–483 (2009)
3. Osentoski, S., et al.: Remote robotic laboratories for learning from demonstration. Int. J. Soc. Robot. **4**(4), 449–461 (2012)
4. Calinon, S., Billard, A.: Incremental learning of gestures by imitation in a humanoid robot. In: Proceedings of the ACM/IEEE International Conference on Human-Robot Interaction. ACM (2007)
5. Sheridan, T.B.: Human–robot interaction status and challenges. Hum. Factors: J. Hum. Factors Ergon. Soc. **58**(4), 525–532 (2016)
6. Gombolay, M.C., Huang, C., Shah, J.A.: Coordination of human-robot teaming with human task preferences. In: AAAI Fall Symposium Series on AI-HRI, vol. 11 (2015)
7. Rautaray, S.S., Agrawal, A.: Vision based hand gesture recognition for human computer interaction: a survey. Artif. Intell. Rev. **43**(1), 1–54 (2015)
8. Schou, C., et al.: Human-robot interface for instructing industrial tasks using kinesthetic teaching. In: 44th International Symposium on Robotics (ISR). IEEE (2013)
9. Vasic, M., Billard, A.: Safety issues in human-robot interactions. In: IEEE International Conference on Robotics and Automation (ICRA). IEEE (2013)
10. http://www.ros.org/. Accessed 10 Apr 2017
11. Qian, K., Niu, J., Yang, H.: Developing a gesture based remote human-robot interaction system using kinect. Int. J. Smart Home **7**(4), 203–208 (2013)
12. Ratner, E., et al.: A web-based infrastructure for recording user demonstrations of mobile manipulation tasks. In: IEEE International Conference on Robotics and Automation (ICRA). IEEE (2015)
13. http://gazebosim.org/. Accessed 10 Apr 2017

14. Di Nuovo, A., et al.: A web based multi-modal interface for elderly users of the robot-era multi-robot services. In: IEEE International Conference on Systems, Man and Cybernetics (SMC). IEEE (2014)
15. Toris, R., Kent, D., Chernova, S.: The robot management system: a framework for conducting human-robot interaction studies through crowdsourcing. J. Hum.-Robot Int. 3(2), 25–49 (2014)
16. Osentoski, S., et al.: Robots as web services: reproducible experimentation and application development using rosjs. In: 2011 IEEE International Conference on Robotics and Automation (ICRA). IEEE (2011)
17. http://gazebosim.org/gzweb. Accessed 10 Apr 2017
18. Doumanoglou, A., Kouskouridas, R., Malassiotis, S., Kim, T.K.: Recovering 6D object pose and predicting next-best-view in the crowd. In: Proceedings of the IEEE Conference on Computer Vision and Pattern Recognition, CVPR (2016)
19. Oikonomidis, I., Kyriazis, N., and Argyros, A.A.: Efficient model-based 3D tracking of hand articulations using Kinect. In: BMVC, vol. 1, no. 2, p. 3 (2011)
20. Piperagkas, G.S., Mariolis, I., Ioannidis, D., Tzovaras, D.: Keyframe extraction with semantic graphs in assembly processes. IEEE Robot. Autom. Lett. PP(99), 1–1 (2017). doi:10.1109/LRA.2017.2662064

Mobile VR Headset Usability Evaluation of 2D and 3D Panoramic Views Captured with Different Cameras

Alessio Regalbuto[✉], Salvatore Livatino, Kieran Edwards, and Iosif Mporas

School of Engineering and Technology, University of Hertfordshire, Hatfield, UK
alessioregalbuto@outlook.com

Abstract. Virtual Reality has for many years been an attractive option for simulation, prototyping and tele-operation. Today new technologies have become available, and Virtual Reality has got new momentum. A new type of VR headset has been proposed based on smartphones. This device opens new application possibilities including remote environments tele-exploration. New cameras have become available, which suit the new VR headsets. In this paper we present an assessment of panoramic views for mobile VR headsets using different types of cameras that have different costs. We present the results of this assessment and discuss the achieved performance when the generated images are observed through mobile VR headsets.

Keywords: Mobile VR · 2D panorama · 3D panorama · Stereo camera · Fish-eye camera · 360 camera

1 Introduction

Recently a growing interest in Virtual Reality applications capable of exploring remote environments is prospecting new possibilities for several industries. Applications such as planetary exploration [1] and robotic control systems for tele-operations [2, 3], started to adopt Virtual Reality to achieve better performances. Beside advantages provided by VR interactive controllers, a crucial role is covered by VR visualization systems.

Several studies, e.g. [4, 5], proposed and assessed the use of stereoscopic 3D visualization (S3D) for robotic tele-operations, demonstrating the advantage of these visualization systems over traditional 2D. Modern visualization systems for panoramic visualization are also inspiring the development of new applications to enhance navigation systems and remote environment explorations.

Today we are also assisting the spread of modern smartphones, which have the potential to operate with Virtual Reality using Mobile VR headsets.

Despite the development of interactive systems to navigate panoramic virtual maps using a smartphone [6], panoramic visualization is still an open issue in terms of panorama acquisition.

© Springer International Publishing AG 2017
A. Ronzhin et al. (Eds.): ICR 2017, LNAI 10459, pp. 191–200, 2017.
DOI: 10.1007/978-3-319-66471-2_21

In this paper, we investigated different techniques available today to produce panoramic views of real places, using four cameras: a smartphone camera using Google Cardboard Camera App (GCCA), a Fuji W3 camera, a Theta S camera, and a Flylink action camera. We also performed a usability evaluation using Mobile VR to analyse performances of generated panoramas in terms of realistic impression, sense of presence, image sharpness, and depth-perception. In the following sections, we discuss five panoramic acquisition methods, and present results of our usability evaluation.

2   Panorama Acquisition and Visualization

In this section, we describe five different panoramic acquisition systems we used for the evaluation. All panoramas were collected and uploaded online to be accessed by an LG G3 mobile phone remotely. We used a Shinecon Mobile VR headset to view both 2D and 3D panoramas.

2.1   3D Panorama Using Google Cardboard Camera App (GCCA)

Google Cardboard Camera App is a mobile application developed by Google that can acquire a cylindrical 3D panorama from a video. The algorithm of the app analyses camera panoramic rotation to extract motion parallax and generate disparity maps of the entire environment. Disparity maps are then used to automatically generate a final stereoscopic stitched panorama image. To visualize the panorama in 3D, the app provides Virtual Reality mode, which needs a Mobile Virtual Reality headset.

In Fig. 1 we show a snapshot of a specific position of the panorama. It should be noted that even if the panorama is not spherical, a blurred color is added by the app to the missing portions of the sphere (Table 1).

Fig. 1. Google Cardboard Camera App – 3D Cylindrical Panorama

Table 1. Advantages and disadvantages of GCCA panorama acquisition

Advantages	Disadvantages

• Easy and fast panorama acquisition
• Multiplatform application
• No need to buy expensive cameras or pan
  tilt units
	• Non-spherical panorama (cylindrical)
	• Computed parallax from a single lens
	  (possible errors)
	• Errors when moving objects in the scene

2.2 3D Panorama Using Fuji 3D W3 Camera

This method was tested using a pan tilt unit and a Fuji W3 Stereo Camera (Fig. 2). Inspired by the technique used by Smith et al. [7], we rotated the camera to capture the complete spherical environment, maintaining camera position fixed to the centre of rotation. A total number of 280 pictures was taken, followed by some image stitching operations (Table 2).

Fig. 2. Fuji W3 Stereo Camera

Fig. 3. Ricoh Theta S Panoramic Full-spherical Camera

Fig. 4. Flylink Action Camera

Table 2. Specifications, advantages and disadvantages of 3D Fuji W3 Stereo Camera panorama acquisition

Camera specifications	Observations
• Sensor resolution: 10.0 Megapixel • Optical sensor type: CCD • Optical sensor size: 1/2.3" • System: TTL contrast detection • Focal length wide: 35 mm • Max aperture wide: 3.70	Advantages: • Higher resolution images • Automatic exposure for HDR panoramas Disadvantages: • Expensive setup, long process • Editing needed to correct colours and distortion

2.3 2D Panorama Using Fuji 3D W3 Camera

This method took advantage of the previous method to produce a 2D panorama. This was possible using the same image for both left and right image. The panorama was then uploaded to an online viewer, which supports VR mode visualization.

2.4   2D Panorama Using Ricoh Theta S Camera

Compared to previous methods, this acquisition was done using a Ricoh Theta S camera (Fig. 3), which uses 2 fish-eye 190° field of view lenses. Image stitching is processed automatically by the camera, taking advantage of the 10° overlap of the pictures captured by each lens (Table 3).

Table 3. Specifications, advantages and disadvantages of Ricoh Theta S Camera panorama acquisition

Camera specifications	Observations
• Number of lenses: 2 • Video resolution: 1920 × 1080 at 30 fps/16 Mbps • Photo resolution: 5376 × 2688 (24 Megapixels) • Aperture: f/2.0	Advantages: • Easy and fast acquisition • Cheap system, automatic stitching • Pan tilt unit not needed Disadvantages: • Less resolution than panoramas created using pan tilt units • Large field of view of the lenses may introduce distortions

2.5   2D Panorama Using Flylink Camera

This acquisition method uses a Flylink action camera (Fig. 4) to take 3 wide angle pictures and stitch them using commercial software. It should be noted that this method requires image rectification to complete the stitching properly (Table 4).

Table 4. Specifications, advantages and disadvantages of Flylink Action Camera panorama acquisition

Camera specifications	Observations
• Lens: 170-degree wide-angle • Video Resolution: 1080P (1920 * 1080) 30FPS, 720P (1280 * 720) 30FPS, VGA (848 * 480) 30FPS, QVGA (640 * 480) 30FPS • Photo: 12 M/8 M/5 M • Video format: MOV • The video coding: H.264	Advantages: • Very cheap system • Less than 5 shots needed for an acceptable panorama Disadvantages: • Low resolution • Distorted image to be rectified and stitched in post-production • Long post processing

3   Usability Evaluation Procedure

According to guidelines indicated by the literature [8–10], we asked twelve participants to take part in our evaluation and complete a specifically designed questionnaire using Google Forms. Our group of test-users consisted of 3 women and 9 men, aged between 20 and 55 years, with varied experience in videogames, 3D movies visualization and

Virtual Reality headsets. Test-users were asked to perform multiple observations of the different panoramas through a mobile VR display.

Each test-user provided an estimate of 4 qualitative parameters: Depth Perception, Presence (sense of presence), Realism (realistic visual impression), and Sharpness (image sharpness). These referred to careful observation of 5 different panoramic views (captured with different cameras) of the same location.

Each parameter was evaluated on each viewed panorama, using a single focused question. Responses were provided using the 7-point Likert scale. Furthermore, users were asked to estimate distances between camera and specific objects on each panorama, and these were compared to ground truth.

We used both descriptive and inferential statistics to process users' response (mean values, standard error, t-Student). The overall score of each parameter on a viewed panorama was estimated using the average of all corresponding responses. Results are shown through tables and diagrams in the next section.

4 Results

In the following we compare the five panorama acquisition methods, analysing obtained scores for Depth Impression, Presence, Realism, Sharpness, and Distance Evaluation. Results are based on multiple paired t-test analysis: the score of each panorama parameter was compared to the corresponding score of the others using t-tests.

4.1 Depth Impression

In Fig. 5 we reported the mean and standard error of the achieved Depth Impression of each panorama. In Table 5, p-values lower than 0.05 are presented, to highlight significant differences between panoramas. P-values higher than 0.05 are ignored.

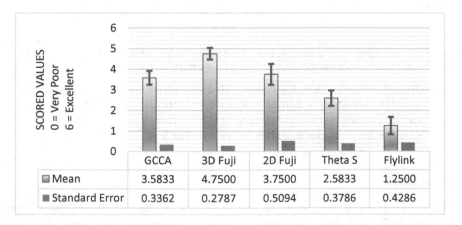

	GCCA	3D Fuji	2D Fuji	Theta S	Flylink
Mean	3.5833	4.7500	3.7500	2.5833	1.2500
Standard Error	0.3362	0.2787	0.5094	0.3786	0.4286

Fig. 5. Statistical comparison of the 5 panorama acquisition methods, in terms of Depth Impression. The higher value implies a better depth perception of the panorama

Table 5. In this table, we present p-values in ascending order, calculated using multiple t-tests on the scored Depth Impression of the panoramas. Values over 0.05 are ignored

3D Fuji – Flylink (p = 0.00015)	2D Fuji – Flylink (p = 0.00051)	Theta S – Flylink (p = 0.00125)	3D Fuji – Theta S (p = 0.00128)
GCCA – Flylink (p = 0.00205)	2D Fuji – Theta S (p = 0.01498)	*GCCA* – 3D Fuji (p = 0.0238)	–

4.2 Presence

In Fig. 6 we reported the mean and standard error of the achieved Presence of each panorama. In Table 6, p-values lower than 0.05 are presented, to highlight significant differences between panoramas. P-values higher than 0.05 are ignored.

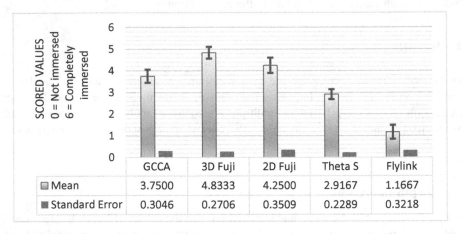

Fig. 6. Statistical comparison of the 5 panorama acquisition methods, in terms of Presence (feeling of being there). The higher value implies a higher level of immersion inside the virtual panorama

Table 6. In this table, we present p-values in ascending order, calculated using multiple t-tests on the scored Presence of the panoramas. Values over 0.05 are ignored

2D Fuji – Flylink (p = 2.00144E-05)	Theta S – Flylink (p = 2.2696E-05)	3D Fuji – Flylink (p = 4.74656E-05)	*GCCA* – 3D Fuji (p = 0.0002)
GCCA – Flylink (p = 0.0009)	3D Fuji – Theta S (p = 0.00138)	2D Fuji – Theta S (p = 0.01039)	–

4.3 Realism

In Fig. 7 we reported the mean and standard error of the scored Realism of each panorama. In Table 7, p-values lower than 0.05 are presented, to highlight significant differences between panoramas. P-values higher than 0.05 are ignored.

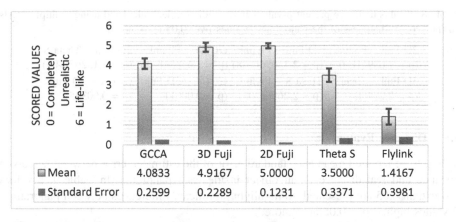

Fig. 7. Statistical comparison of the 5 panorama acquisition methods, in terms of Realism. The higher value implies a more realistic perceived panorama

Table 7. In this table, we present p-values in ascending order, calculated using multiple t-tests on the scored Realism of the panoramas. Values over 0.05 are ignored

2D Fuji - Flylink	3D Fuji - Flylink	*GCCA* - Flylink	2D Fuji - Theta S
(p = 1.8728E-07)	(p = 1.25015E-05)	(p = 2.01588E-05)	(p = 0.0003)
Theta S - Flylink	*GCCA* - 3D Fuji	*GCCA* - 2D Fuji	3D Fuji - Theta S
(p = 0.00056)	(p = 0.00201)	(p = 0.00207)	(p = 0.00224)

4.4   Sharpness

In Fig. 8 we reported the mean and standard error of the achieved Sharpness of each panorama. In Table 8, p-values lower than 0.05 are presented, to highlight significant differences between panoramas. P-values higher than 0.05 are ignored.

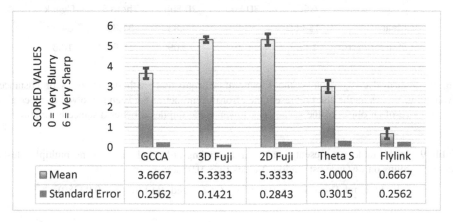

Fig. 8. Statistical comparison of the 5 panorama acquisition methods, in terms of Sharpness. The higher value implies a sharper perceived panorama

Table 8. In this table, we present p-values in ascending order, calculated using multiple t-tests on the scored Sharpness of the panoramas. Values over 0.05 are ignored

2D Fuji – Flylink	3D Fuji – Flylink	GCCA – Flylink	2D Fuji – Theta S
(p = 1.08911E-08)	(p = 2.35076E-08)	(p = 1.66854E-06)	(p = 1.15267E-05)
3D Fuji – Theta S	Theta S – Flylink	GCCA – 3D Fuji	GCCA – 2D Fuji
(p = 2.2696E-05)	(p = 2.2696E-05)	(p = 0.00011)	(p = 0.00345)

4.5 Distance Evaluation

In Fig. 9 we reported mean and standard error of the estimated error between perceived distance and real distance of an object inside each panorama. In Table 9, p-values lower than 0.05 are presented, to highlight significant differences between panoramas. P-values higher than 0.05 are ignored.

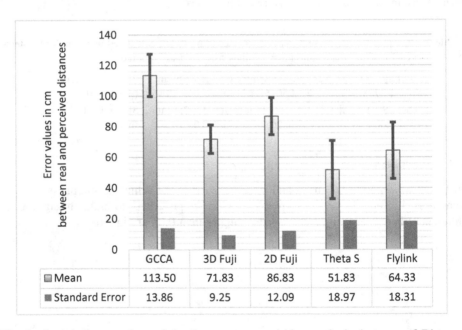

	GCCA	3D Fuji	2D Fuji	Theta S	Flylink
Mean	113.50	71.83	86.83	51.83	64.33
Standard Error	13.86	9.25	12.09	18.97	18.31

Fig. 9. Statistical comparison of the 5 panorama acquisition methods, in terms of Distance Evaluation. Values on the chart represent the error in centimetres between perceived distance and real distance of an object of the scene. The lower value implies a better distance estimation

Table 9. In this table, we present p-values in ascending order, calculated using multiple t-tests between groups on distance estimation. Values over 0.05 are ignored

2D Fuji and Theta S (p = 0.01828)	GCCA and 3D Fuji (p = 0.03380)

5 Discussion

According to results, we can conclude that panoramic acquisitions using Fuji 3D W3 camera achieved best scores in terms of Depth Impression, Presence, Realism, and Sharpness. Furthermore, there is no significant difference between 2D and 3D panoramas created using the Fuji camera. This is an unexpected result especially for Depth Impression: we expected the 3D panorama to provide higher depth compared to 2D panorama. We believe this result is due to several monocular depth cues that were captured inside the scene, inducing depth perception on the viewer even without binocular depth cues.

Another interesting outcome is that even if GCCA showed significant lower scores compared to 3D Fuji camera for Depth Impression, Presence, Realism, and Sharpness, it obtained significant better scores for Distance Evaluation.

Results also proved that Flylink action camera is the less performant panorama acquisition method compared to all the others, except for Distance Evaluation where Theta S camera resulted equally performant.

Depending on the application and on time requirements for the final stitched panorama generation, the following guidelines are provided:

Time-Constrained Applications: The best option to acquire an acceptable panorama in very short times is the Theta S acquisition method. Using this camera, the panorama will be generated in less than 10 s. According to Theta S results, the performance for Depth Impression, Sharpness, Presence and Realism is satisfactory. GCCA and Fuji methods provide higher quality compared to Theta S, but spending much longer times, especially in the case of the Fuji camera.

An alternative to the Theta S is the GCCA, which performs better spending longer times, but with no need of post processing corrections.

No Time-Constrained Applications: For applications that do not have strict time-constraints the Fuji camera acquisition method represents the best option to achieve highest performance in terms of Realism, Presence and Sharpness. Alternatively, if post processing panorama stitch is not desirable, similar performance can be achieved with the GCCA acquisition method.

Distance Evaluation Applications: When accurate distance evaluation is required, our results suggest adopting the GCCA acquisition method. This performs significantly better than the 3D Fuji camera. However, if Realism and Presence are equally important, the Fuji camera can be a valid alternative compared to Theta S and Flylink acquisition methods.

6 Conclusion

This paper assessed five different panoramic acquisition methods for VR headsets based on smartphones, using different cameras. The Google Cardboard Camera App, 3D Fuji camera, 2D Fuji camera, Ricoh Theta S camera and Flylink action camera were compared and evaluated. The results provided guidelines to help choose the best acquisition method according to application requirements. This paper supports the development of better panoramic images for Mobile VR applications, for applications aimed at realistic tele-presence, remote environments explorations and tele-operation.

References

1. McGreevy, M.W.: Virtual reality and planetary exploration. Virtual Reality. Appl. Explor. 46–65 (1993)
2. Tang, X., Yamada, H.: Tele-operation construction robot control system with virtual reality technology. Procedia Eng. **15**, 1071–1076 (2011)
3. Kato, Y.: A remote navigation system for a simple tele-presence robot with virtual reality. In: IEEE International Conference on Intelligent Robots and Systems, pp. 4524–4529 (2015). https://doi.org/10.1109/IROS.2015.7354020
4. Livatino, S., Muscato, G., Sessa, S., Neri, V.: Depth-enhanced mobile robot teleguide based on laser images. Mechatronics **20**(7), 739–750 (2010)
5. Livatino, S., Muscato, G., De Tommaso, D., Macaluso, M.: Augmented reality stereoscopic visualization for intuitive robot teleguide. In: Proceedings of IEEE ISIE, pp. 2828–2833 (2010)
6. Wagner, D., Mulloni, A., Langlotz, T., Schmalstieg, D.: Real-time panoramic mapping and tracking on mobile phones. In: Proceedings – IEEE Virtual Reality, pp. 211–218 (2010). https://doi.org/10.1109/VR.2010.5444786
7. Smith, N.G., Cutchin, S., Kooima, R., Ainsworth, R.A., Sandin, D.J., Schulze, J., DeFanti, T.A.: Cultural heritage omni-stereo panoramas for immersive cultural analytics – from the Nile to the Hijaz. In: 8th International Symposium on Image and Signal Processing and Analysis (ISPA), pp. 552–557 (2013)
8. Steuer, J.: Defining virtual reality: dimensions determining telepresence. J. Commun. **42**(4), 73–93 (1992)
9. Nielson, J.: Usability Engineering. Morgan Kaufmann, San Mateo (1993)
10. Livatino, S., Koeffel, C.: Simple Guidelines for Testing VR Applications. In: Advances in Human Computer Interaction. InTech, 600 p. (2008)

Crossover of Affective Artificial Intelligence Between Robotics Simulator and Games

Hooman Samani[1]([✉]) and Doros Polydorou[2]

[1] Department of Electrical Engineering, National Taipei University,
New Taipei, Taiwan
hooman@mail.ntpu.edu.tw
[2] University of Hertfordshire, Hertfordshire, UK
d.polydorou@herts.ac.uk

Abstract. The aim of this paper is to share state of the art in the field of robotics and artificial intelligence from one side and gaming from other side. Inspired from affective computing in social robotics and in order to improve the learning factor of serious games, we introduce an affection layer which improves the emotional intelligence of a game character. As the components of that layer we propose modules which can be integrated in a game engine in order to enhance the verisimilitude of the virtual world. The proposed architecture can be integrated in several games to improve their emotional abilities which can lead to developing believable characters in the game environment. We believe that such ability increase the motivation for the user to learn since possibilities and situations would be much pragmatic.

Keywords: Interactive collaborative robotics · Games · Simulation · Artificial intelligence

1 Introduction

In this paper, we propose a layer in game engines that could provide more realistic behavior in the serious game environment which can influence the improvement of learning directly. That aspect has been improved in robotics field, especially with social robots. Hence such technology can be employed in games.

There are many (serious) games which deal with social issues which try to promote empathy in a player. Social interactive games such as Marriage [1] and Sims [2] let players interact with virtual characters [3, 4]. Games in the market are now becoming more and more emotionally mature aiming at connecting with the player [5–7].

The new trend in games is to try and inject emotions into games [8]. The lack of emotional attachment in gaming is a notion widely accepted. Few games, if any, manage to make the player feel empathy for the character that they are controlling or sympathy for the characters around him. Even though that can be accounted to a number of reasons, a big setback is the fact that virtual worlds do not feel real. The characters which share this world with the player are lifeless, with prefixed static behaviors and actions. An experienced player knows and has grown accustomed to expect just that. NPC's are there either to provide support, information, give guests or

© Springer International Publishing AG 2017
A. Ronzhin et al. (Eds.): ICR 2017, LNAI 10459, pp. 201–208, 2017.
DOI: 10.1007/978-3-319-66471-2_22

sell items. For example, no matter how many times you ask them the same thing or how badly you treat them; they will still be there to do their assigned task. This expectation automatically lowers the believability of the characters and the world, making the player less concerned about the consequences of his/her behavior, thus subsequently not caring about the players they are meant to feel empathy and sympathy for. In order for games to be taken seriously and used as a tool for education, this problem must be remedied. Play is defined as the navigation of a suite of choices (i.e. decisions), where each decision leads to an action that has a discernable outcome. Currently, the choices that a player makes in a virtual world are reflected in results which are nonrealistic but limited by the Artificial Intelligence of the game. This kind of suspension of disbelief might be acceptable for commercial games but not for serious games, especially if their creators aim to get them approved as a successful learning medium. Creating a world where the people in it have feelings and act according to them can motivate and relate to their emotional Intelligence. Emotional Intelligence is defined [9] as "the ability to monitor one's own and others' feelings and emotions, to discriminate among them and to use this information to guide one's thinking and actions". This paper argues that Emotional Intelligence can have a significant influence on how successful serious games can be as a learning tool.

2   Implementation of Emotional Intelligence in Serious Games

The emotional intelligence platform is a system that aims to make a virtual world more believable by simulating human emotions and applying them to NPC's. Looking specifically at Serious Games, the platform can be applied as a base in the creation of a virtual environment. Emotions will play an integral part of how the world functions and offer a great sense of unpredictability to it. This will make the virtual worlds more believable for the player, offer more accurate simulated results and even oppose the general misconception that games are just for fun and can offer nothing more than that. There is currently a lot of research in emotion theory. One notable example is the Gamegdala engine [10], which informs a lot of our methodology.

Our work assumes that each NPC agent has two basic characteristics: goals and beliefs. Furthermore, there is a great emphasis on social interactions between the characters and that is the main method of changing influencing these beliefs as it is explained in the example below:

2.1   User Scenario

Each NPC is assigned with certain random characteristics and an area of influence. As the player explores the world and interacts with the NPC's these characteristics change according to the actions of the player, whenever he/she is in their area of influence. Furthermore when that NPC whose characteristics changed enters the area of influence of another NPC those player influences are transferred over to second NPC as well. To make things easier to understand, let's take a simple example. Let's imagine a game that aims to teach the culture of a country. The player is free to roam in a city market and enters a shop. In the shop there are two NPCs, a female and the shop owner. The player chats politely with the shop owner. The female NPC is in the vicinity of the conversation, therefore because the player was polite, her likability towards the player increased as well. The female NPC then moves away from the shop, entering a second shop. Since she will enter the area of influence of the shop owner of the second shop, the likability influences towards the player are transferred to the second shop owner as well. If the player now enters the second shop, since the shop owner already "heard" positive comments from the female NPC about him, he can offer him a small discount or different dialogue options. We are of course making the assumption here, that the second shop owner has the character trait of trusting people easily and the female NPC has the tendency to talk openly to everyone. Since in the beginning all these character traits were distributed randomly nobody can predict how the situation will play out. Depending on how the player treats the NPCs, there will be a different reaction. The player by realizing that, will no longer treat the people in the world with contempt and will start thinking about their feelings as well. This way the player can learn the importance of good behavior and politeness. Continuing on the example of the cultural game, let us now assume that we want to use the same platform for a different country. In this country the people are less open and more private with their lives. By adjusting some basic values, we can easily control the original random character trait assignment to reflect the change. Therefore now the engine will be more conservative on the values it will assign to each NPC. The aim of this system is to enhance the verisimilitude of the virtual world. Crucial game play objectives which are vital for the continuation of the story will of course not be left at random. They will still be scripted in the world by the game designer. Compared to older systems the Emotional Intelligence Platform can also offer new creative perspectives without the need of scripting everything. For example let us imagine that the objective of the player is to get inside a locked door. The guard standing outside has the keys and in order to let the player in he must have a high trust value towards the player. In order for that trust to be raised, the guard asks the player to perform a task for him. If however, the player did something influential to an NPC who has access to the guard, then the trust level towards the player might already be high, thus offering to the player another way in without having to complete the task. Since the influences platform is already set up, the game designer will only have to script the behavior of the guard when the trust level is high enough, saving valuable time from having to script the event from the beginning. Furthermore, this platform can educate players on the importance of honesty, trust and friendship and show that every action can have an effect on the world around you.

3   Development of Emotional Intelligence

Our proposal is that a character in a game should be equipped with an intelligent emotional module which controls the affective behavior of character in the game. Emotional properties are mathematically modeled to generate a platform for overall feelings in characters. That model would control the affective state of the character. In this section we describe the 3D emotional space and the mechanism for transition in that system. We also explain the possibility of involving artificial endocrine system in that emotional module. We consider interaction area around each agent within the game engine. When two agents locate in the interactive zone, emotional expression values of each agent would be transferred to the corresponding agent.

3.1 Emotional Modeling in the Game Engine

The internal emotional property of a character in a game can be modeled as an affective state system. In this section we describe an affective state model and propose a systematic method for handling changes in such model.

3.1.1 Affective State

Besides realizing the emotional expressions of the interaction, the system may develop the internal state of the character to handle the overall emotional situation.

We present the affective state model in the three dimensional space and then describe a systematic method to demonstrate the changes in affective states of the agent. That comprehensive model provides a complete platform for the emotional state of the character by considering all mixed emotions. Furthermore, a realistic transition system is proposed to control changes in internal affective states which have not been investigated in previous emotional models.

3.1.2 Modeling the Affective State

Tension and energy are believed as two principle parameters for representing the mood of human being in psychological studies [11]. We have considered these two dimensions as Activation (act) and Motivation (mot) axes in the 2D affective state plane in order to categorize affects in a methodical manner.

Affective state areas can be modeled by bean-shaped curves of genus zero with a single singularity with an ordinary triple point at the origin to illustrate their coverage in the affective coordinate system:

$$(x^2 + y^2)2 = x^3 + y^3 + a(x^2 + x - y),$$ (1)

which gives a crooked egg curve when a is zero, and a bean curve when a is one.

State areas can be modeled by transecting the origin for plotting the above closed curve as $X_i = x_i - x_{act_i}$, $Y_i = y_i - y_{mot_i}$ in the Activation-Motivation plane. So any affective states can be identified according to its transferred origin as Eq. 2

$$O_s = \left[X_i = x_i - x_{act_i}, Y_i = y_i - y_{mot_i}, Z_j^i = z_{Sub_j^i} \right] \text{ where } \{1 \leq i \leq 25, 1 \leq j \leq 10\},$$ (2)

where i represents state and j shows the sub-state number.

By having the position of the origins, Eq. 1 could utilize in order to specify affective state areas.

Any of the main affective states include several sub-states which represent more details internal state. Sub-states can be represented as the third dimension of the affective state coordinate.

3.1.3 Affective State Transition

To model the system to link interaction and affective state, the transition in the affective state space has formulated as following:

$$\vec{S}_{t_{Act-Mot-Sub}} = \vec{S}_{t-h} + \eta\vec{\Phi} + \beta\Gamma\vec{\Delta}, \tag{3}$$

where $\vec{S}_t$ is the affective state of the agent in the affective space at time t and $\vec{S}_{t-h}$ is affective state in time $t - h$, where h is the processing time gap as discrete system;

$\vec{\Phi}$ is the vector field over the states which converges to the certain point in affective state coordinate system. $\vec{\Phi}$ can be considered as the gravitational field of a point mass due to a point mass c located at point P_0 having position r_0 as:

$$\Phi = \frac{-kc}{\left|\vec{r} - \vec{r_0}\right|}(\vec{r} - \vec{r_0}), \tag{4}$$

$c > 0$ is a constant, Φ points toward the point r_0 and has magnitude $|\Phi| = \frac{-kc}{|\vec{r}-\vec{r_0}|^2}$;

η is the adjusting parameter for converge vector field;

β is the affective state coefficient which represents the personality of the agent that controls the rate of change in the mood. Larger β means that the state would change faster which makes the agent more moody;

$\vec{\Gamma}$ is the learning rate. The change in state is different when agent has more interaction and this parameter helps to have more realistic changes in affective state;

$\vec{\Delta} = \vec{\Delta}_{Act-Mot-Sub}$ is the 3D normal vector to transfers the state over time in affective state space based on the emotional input according to the interactions. First two components are in the Activation - Motivation plane which are driven from emotional input:

$$\vec{\Delta}_{Act-Mot} = \sum_{m=1}^{6} e_{Mot_m}Mot(i) + \sum_{m=1}^{6} e_{Act_m}Act(j), \tag{5}$$

where Mot and Act are Motivation and Activation axes and e_is are 6 values of happiness, sadness, disgust, surprise, anger, and fear in Activation and Motivation directions.

The third component of Δ represents the movement in sub-state direction which obtained from the rate of the first two components:

$$\vec{\Delta}_{Sub} = \left|\frac{d}{dt}\Delta_{Act-Mot}\right|(k). \tag{6}$$

In this way the vector $k\,\Gamma\,\Delta$ finds its direction to reach the next affective state.

3.1.4 Artificial Endocrine System

Natural endocrine system is viewed as a network of glands that works with the nervous system to secrete hormones directly into the blood so as to control the activity of internal organs and coordinate the long range response to external stimuli. Hormones which are chemicals released by components of the endocrine system affect other parts of the body and play a significant role in the endocrine system so as to preserve homeostasis. Here, we will introduce the relation of hormones with human emotion and behavior which can be implemented into the emotional layer of the agent in the game.

Virtual biological systems considered as research field of biological inspired computing. Artificial Neural Network (ANN) is one of the well-known tools in computational intelligence techniques. In the same way, the endocrine system could also be a very useful tool.

This paper focuses on the hormones which are related to emotions and biological qualities [12].

In our system all hormones are considered to be secreted by two parameters:

- The activation function which can be presented by employing the logistic function and
- The gland bustle that should be considered through all the stimuli channels.

So the glands secretion can be modeled as Eq. 7:

$$\Lambda_q = \frac{1}{1 + \exp(-aq)} \sum_{q=1}^{m \times m} p_i \Theta_q. \tag{7}$$

Above representing shows that the gland secretion, Λ, is the product of the each gland bustle, Θ_q, by considering ρ_q as the stimuli weight, which can be activated through the nonlinear activation function $\frac{1}{1 + \exp(-aq)}$. The gland bustle should be considered over m emotional values. n is number of different agents in the game environment that considered agent has interaction with them. The coefficient a in the activation function depends on the current volume of the hormone in the system.

4 Experimental Results

We have developed a simulator and applied the mentioned theories to that virtual environment. We designed the emotional layer for that system. The overall platform of this simulator is presented in Fig. 1. At the beginning of the game, 10 agents are placed in random positions in the environment. For each of them, random interaction area is considered as well. Agents generate random emotional values in their interaction area. The user can start the game by navigating through the environment using mouse clicks. When interaction areas of two agents meet, the emotional values are transferring between those agents and according to mentioned theories, the affective state of each one changes. The levels of hormones also change by time base on situations and interaction.

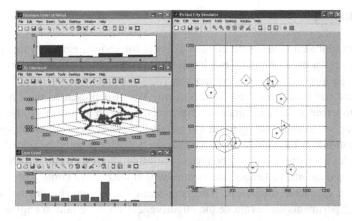

Fig. 1. The game simulator environment

We focus on affective properties of the character in the game and introduce the systematic method in order to generate such affection. The affective state system also developed as it has described in the previous sections. Figure 2 presents the affective state space which includes affective sates that are generated according to the Eq. 1, by using bean shape closed curves. Each of affective sub-states are demonstrated by volume in the 3D space and, as illustrated, several affective states overlap and share the subspace in the affective system which demonstrates the mixture of the emotional states. The state of the agent can be considered as one point inside this space. If the position of that point confined inside any of these volumes, then the internal state of the agent belongs to that category of affective state.

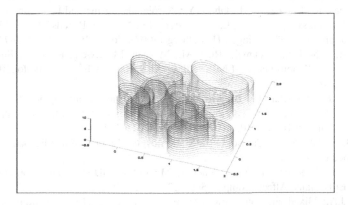

Fig. 2. Affective states are modeled as volumes in the 3D affective coordinate system

5 Conclusion

We have introduced an intelligent emotional layer for serious games in this paper. We believe that such a module can be employed in several serious game platforms in order to benefit learning aspects. A comprehensive emotional model which considers affective state has been presented for the agent. We also presented the system for an artificial endocrine system in order to improve the affective power of the agent. By implementing such emotional agent in a simulator, we have developed the system with realistic emotional behaviors. We believe that such platform could be added to several games in order to improve realistic behavior. Building about the Gamygdala engine, we have introduced an artificial endocrine system which we believe could replicate an even more realistic character simulation. The next steps for this work, will be to conduct even more experiments and evaluate the effectiveness on the emotion engine into more scenarios.

References

1. The Marriage. http://www.rodvik.com/rodgames/marriage.html. Accessed 11 July 2017
2. Sims. https://www.thesims.com/. Accessed 11 July 2017
3. Dungeons and Dragons. http://dnd.wizards.com/. Accessed 11 July 2017
4. Davies, M.: Designing Character-Based Console Games. Charles River Media, Boston (2007)
5. Bosser, A.-G., Levieux, G., Sehaba, K., Buendia, A., Corruble, V., Fondaumière, G., Gal, V., Natkin, S., Sabouret, N.: Dialogs taking into account experience, emotions and personality. In: Ma, L., Rauterberg, M., Nakatsu, R. (eds.) ICEC 2007. LNCS, vol. 4740, pp. 356–362. Springer, Heidelberg (2007). doi:10.1007/978-3-540-74873-1_42
6. Thawonmas, R., Kamozaki, M., Ohno, Y.: A role casting method based on emotions in a story generation system. In: Ma, L., Rauterberg, M., Nakatsu, R. (eds.) ICEC 2007. LNCS, vol. 4740, pp. 182–192. Springer, Heidelberg (2007). doi:10.1007/978-3-540-74873-1_22
7. Dörner, R., Göbel, S., Kickmeier-Rust, M., Masuch, M., Zweig, K. (eds.): Entertainment Computing and Serious Games. LNCS, vol. 9970. Springer, Cham (2016). doi:10.1007/978-3-319-46152-6
8. Broekens, J., Hudlicka, E., Bidarra, R.: Emotional appraisal engines for games. In: Karpouzis, K., Yannakakis, Georgios N. (eds.) Emotion in Games. SC, vol. 4, pp. 215–232. Springer, Cham (2016). doi:10.1007/978-3-319-41316-7_13
9. Mayer, J.D., Salovey, P., Caruso, D.R., Sitarenios, G.: Emotional intelligence as a standard intelligence (2001)
10. Popescu, A., Broekens, J., Van Someren, M.: GAMYGDALA: an emotion engine for games. IEEE Trans. Affect. Comput. 5(1), 32–44 (2014)
11. Russell, J.A.: Mixed emotions viewed from the psychological constructionist perspective. Emot. Rev. 9(2), 111–117 (2017)
12. Samani, H., Saadatian, E., Jalaeian, B.: Biologically inspired artificial endocrine system for human computer interaction. In: Kurosu, M. (ed.) HCI 2015. LNCS, vol. 9169, pp. 71–81. Springer, Cham (2015). doi:10.1007/978-3-319-20901-2_7

Animal Robot as Augmentative Strategy to Elevate Mood: A Preliminary Study for Post-stroke Depression

Syamimi Shamsuddin[1](✉), Winal Zikril Zulkifli[1], Lim Thiam Hwee[2], and Hanafiah Yussof[3]

[1] Faculty of Manufacturing Engineering,
Universiti Teknikal Malaysia Melaka, 76100 Melaka, Malaysia
syamimi@utem.edu.my
[2] Psychosocial Department, SOCSO Tun Razak Rehabilitation Center,
75450 Melaka, Malaysia
lim.thiamhwee@rehabmalaysia.com
[3] Faculty of Mechanical Engineering,
Center for Humanoid Robots and Bio-Sensing (HuRoBs),
Universiti Teknologi MARA, Shah Alam, Malaysia

Abstract. Animal-assisted therapy is a widely recognized therapy to improve mood in treating depression. In Malaysia, the prevalence of depression is estimated between 8–10% of the population. This study is the first one in Malaysia to use PARO; an animal robot as adjunct therapy for a patient with post-stroke depression. Earlier studies show that PARO helps to increase mood, make people happy and encourages human-to-human interaction. The aim of this study is to introduce PARO as a short-term companion to help patient manage depression during rehabilitation period at a multidisciplinary center. Patient was exposed to PARO for 10 min every day for 3 consecutive days. Results show that PARO helped the patient to manage her psychological distress. Her mood improved and she expressed more smiles when holding PARO. This study suggests that PARO effectively uplifts mood and helps patient to be calm. Further research is warranted on the use of PARO for more patients affected with depression.

Keywords: Human-robot interaction · Robot-assisted therapy · Animal robot · PARO · Depression

1 Introduction

Mental healthcare with the aid of robots is fast becoming an exciting niche in human-robot interaction (HRI) studies. Recent studies have covered the role of robots in clinical settings and care facilities for people with dementia, Alzheimer and also children with special needs. Robot-based solutions are desirable as they are specific, repetitive, motivating and can be made to fit current therapy models of the target population.

© Springer International Publishing AG 2017
A. Ronzhin et al. (Eds.): ICR 2017, LNAI 10459, pp. 209–218, 2017.
DOI: 10.1007/978-3-319-66471-2_23

1.1 Post-stroke Depression

Depression is present in 25–30% among stroke patients [1]. Stroke survivors may experience stress, worry, sadness and hopelessness in different degrees. They may also experience low mood, feelings of hopelessness, withdrawal from daily social activities and even suicide. Several factors that could accelerate symptoms of depression in a stroke patient are lack of psychosocial support, inadequate of nutrients, intrapersonal coping skills and society integration. The dearth of goal-oriented behaviour that guides the patient towards community integration may perpetuates it further.

Post-stroke depression has negative effects on functional recovery [2]. Thus it needs to be treated effectively. Existing treatments include anti-depressant nortriptyline (a type of drug), social care through peer support group and animal-assisted therapy. Anti-depressant helps to improve mood. Social participation and integration reduces emotional distress and improves functional independence. However, community integration may only start after inpatient rehabilitation process took place. Animal-assisted therapy (AAT) is a helpful addition to a rehabilitation regimen. AAT helps motivate and make therapy more enjoyable and less stressful for stroke patients [3]. Nevertheless, the use of animals for therapy is unsuitable for patients whom are afraid of animals or have allergies. Also, the maintenance care will be costly plus the risk of zoonotic infection.

1.2 Animal Robots as Therapy Medium

The creation of robots which is capable to interact with human being and can show facial expression has been popular in the history of robotic [4]. Animal robots is suggested to replace real animals in AAT programs because it can prevent the case of scratches, bite, and allergy but still remain effective and give positive impact. Most robotic animals are modelled after common animals such as dogs and cats. Yet, their behaviour can be unsuccessful in meeting human's anticipations. PARO the seal robot (Fig. 1) has recorded much success to increase mood, make people happy and encourages human-to-human interaction [5].

Fig. 1. PARO the baby seal robot

1.3 Multi-disciplinary Rehabilitation Centre

When a stroke, brain injury or other traumatic event requires that patients spend time in rehabilitation, a working multidisciplinary team is fundamental in delivering effective care across the journey to recovery [6]. Instead of a hospital gown, patients wear their own clothes, and incorporate taking care of themselves as a part of the rehab experience. The amount and types of therapy are prescribed by experts in the rehabilitation unit to help patients set goals and regain the same level of function before their injury.

SOCSO Tun Razak Rehabilitation Center (TRRC) in Melaka is a rehabilitation complex that combines medical and vocational rehabilitation with an allied health institute. Under the SOCSO's 'Return to Work' program, disabled patients undergo physical and vocational rehabilitation in order to rejoin the workforce. Patients in rehabilitation centres need constant mind conditioning and motivation. However, psychologists can only spend a limited time with the patients in a week. Thus, the aim of this pilot study is to propose a robotic animal as an adjunct tool to help provide mental support to the patients and increases mood during the in-patient rehabilitation period.

2   PARO the Baby Seal Robot

2.1   PARO Technical Specifications and Behaviour

PARO is a seal robot designed and developed by Professor Takanori Shibata in 1993. It weighs around 2.8 kg and equipped with several sensors and actuators. PARO features a soft furry coat with built-in intelligence providing psychological, physiological and social effects through physical interaction with humans. It reacts to petting and stroking (Fig. 2) by blinking its eyes and moving its flipper. It cries out when handled roughly, but it reacts positively to soft petting. It acts as a surrogate for a real pet to help reduce stress levels and improve social skills of patients [7]. PARO is very cute and can give facial expressions through the movement of its eyelids [8]. Its battery lasts for one hour and can be charged with a charger resembling a pacifier.

Fig. 2.   PARO being stroked and hugged by a patient at the psychosocial department of TRRC

PARO's model is inspired after a baby harp seal to increase interest among users. Unfamiliar animals are hypothesized to have minimum expectations in the human agent during interaction [4]. Non-technical people (i.e. psychologists, therapists) are receptive to robots as assistive technology; however they require systems that require low technical skills (like programming). PARO is a robot readily equipped with pro-cessors and tactile sensors making it to be directly integrated into current therapy regime.

2.2 PARO's Role to Help Manage Depression

Antidepressants are effective psychological interventions and are easily available in Malaysia [9]. Nevertheless, drugs only act as mood stabilizer and do not cure the illness. These drugs sometimes cause insomnia, fatigue and blurred vision. Patients with depression at centres like TRRC undergo rehabilitation treatment before they are declared stable. Then they can return to their family. In this study, PARO is proposed as an augmentative strategy to elevate mood for a patient with depression during the early stages of the rehabilitation journey after stroke occurred. Early treatment is vital as it helps to decrease the morbidity and mortality associated with depression [10].

3 Robot as Therapy Medium for Post-stroke Depression

3.1 Experimental Protocol

The experimental procedure serves as platform to investigate the interaction between the patient and PARO and to compare the changes before and after robotic exposure. The overall experimental flow is shown in Fig. 3. First, approval was obtained from the UTeM Research Ethics Committee to safeguard the wellbeing of the patients and researchers involved and to ensure that their rights are always protected. Then, suitable patients at TRRC were selected. The inclusion criteria are: diagnosed with depression after stroke, currently an in-patient at the centre, able to converse in Malay language or English, able to stroke an animal and can hear clearly. At the time of study, only one patient fulfils the said criteria at TRRC. Then, a short session with psychologist was conducted to explain about the experiment flow. Another aspect that we have taken into consideration is the "Conditions to Abort Procedures" which include conditions where:

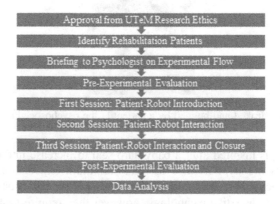

Fig. 3. Overall flow of methodology for the study

a. The patient becomes restless/uncooperative/show abnormal behaviour.
b. The patient is scared of PARO.
c. The patient requests to abort the interaction.

The next step is the pre-experimental evaluation by a rehabilitation psychologist at SRC using the Hamilton Depression Rating Scale (HAM-D), Hamilton Anxiety Rating Scale (HAM-A), Columbia-suicide Severity Rating Scale (C-SSRS) and Pittsburgh Sleep Quality Assessment (PSQI). HAM-D is a validated screening instrument for post-stroke depression [11] to provide indication of depression, guilt, suicide, insomnia (sleep problem) and as guide to evaluate recovery. HAM-A measures the severity of anxiety symptoms and is widely used in both clinical and research settings. C-SSRS rates suicidal assessment on a scale. It is a semi constructed, rater-based interview to prospectively asses the severity and frequency of suicidal ideation and behaviors [12]. PSQI measures the quality and patterns of sleep in adults.

Then, the first interaction with PARO took place. Each session lasted for 20 min every day for 3 consecutive days. During the interaction, the patient was given time to familiarize and was encouraged to interact with PARO under a psychologist's supervision. In the post-experimental stage after the third session, the same four screening instruments: HAM-D, HAM-A, C-SSRS and PSQI were administered. Recorded videos were analysed using OpenCV software. With facial emotion detection, algorithms detect faces and sense micro expressions of smile. The number of smiles were recorded and stored as indication of positive emotions.

3.2   Subject

The single subject in this pilot study was a 43-year old female patient. Before admission as an in-patient at TRRC, she was diagnosed with hemorrhagic stroke in February 2016. When she was in TRRC, she displayed symptoms of depression in which she was referred to psychiatric department in the general hospital of Melaka. Psychiatric findings indicated that the patient has major depressive disorder (MDD). She was prescribed with anti-depressant to stabilize her depression symptoms; to help her sleep better. However, the anti-depressant does not cure depression. It enables patient to return back to previous functioning level. It also does not trigger patients to smile.

3.3   Method of Interaction

Detailed experimental steps using PARO in the therapy session were designed based on advice from the psychologist. The patient was currently undergoing a rehabilitation program that mainly includes physical therapy activities. Three days of therapy with PARO in a week shall complement and not affect the current program. First, the therapy room in TRRC were set-up with HD video camera, voice recorder and projector as shown in Fig. 4. Next, the patient was invited to enter the room. Before the first interaction session, the psychologist administered the four screening instruments. Then, she was given brief introduction about the experiment, her consent was taken and a short video of PARO was shown.

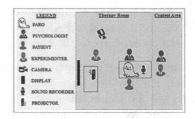

Fig. 4. Room set-up in the rehabilitation centre where the experiment was conducted (left) and layout of therapy room arrangement. Control area only applicable in the third session (right)

Then only the 10-min session can take place to obtain recording of the session without PARO. During this period, the psychologist began to introduce general topics to talk with the patient such as her favourite food, her activities during weekends and her family members. Indirectly, the psychologist was using cognitive behaviour techniques and goal directed behaviour during the therapy. Then, interaction with PARO began. PARO was brought into the room by the experimenter and was put on the patient's lap as she sat in her wheelchair. With PARO in the room, the consultation therapy continues with PARO as companion to the patient. The psychologist continued to have conversations with the patient. The 10-min therapy without PARO and 10-min with PARO cycle was repeated for all three sessions. At the end of each session, the patient was informed that PARO will be taken away. During the third session, before the therapy with PARO ended, the patient was left alone with PARO to observe her behaviour towards the animal robot without human intervention. Patient was observed by the psychologist and experimenter from the next room equipped with a two-way mirror (Fig. 4). Duration of was limited to 20 min because post-stroke patients cannot withstand long therapy sessions and can easily get tired.

3.4   Video Processing Using OpenCV

The sessions without PARO and with PARO were recorded for analysis purposes. To evaluate the patient's positive emotion, specifically the number of smiles, recorded videos were processed using OpenCV software. OpenCV is an open source computer vision library for commercial and research use [13]. It is one of the most widely used libraries in image processing, it is simple to use and has extensive user network.

For smile detection algorithm, Adaptive Boosting (AdaBoost) was used as learning algorithm [14]. Using AdaBoost improves the accuracy of learning algorithm where the output of multiple "weak classifiers" is combined into a weighted sum that represents the final output of the boosted classifier. Smile detection was targeted at the mouth region (Fig. 5). In the smile detection mechanism, the videos were processed frame by frame and converted into grey scale. Smile detection determines 'where' in a face image the 'smile' is located and this is done by scanning the different face image scales and extracting the exact patterns to detect smile. Classifier uses a single feature to define images as smiles or non-smiles which are stored in cascade data.

Fig. 5. Mouth region for image processing (left) and snapshots during therapy sessions (right)

4   Results and Discussion

The first session of the patient-robot interaction was focused on getting the patient to familiarize herself with PARO. Snapshots from the recorded videos are shown in Fig. 5. Based on the video record of the first session, during the first 10 min, the patient responded to the psychologist about general topics and about her family. It was observed that the patient really missed her family. She looked sad and cried when talking about her family. Then, PARO showed up for the first time. The patient got curious about the robot. She stopped crying and the intensity of her previous sad emotions had gone down. When stroking PARO, she kept saying that PARO was cute. With the presence of PARO, it is observed that the patient's mood had changed. PARO reacted to petting and stroking by the patient by blinking its eyes and moving its flipper. This, in turn, continuously prompted the patient to give respond to PARO.

The following day, the session continues with 10 min of session introduction without PARO and 10 min of robot interaction with PARO. The second session started with the discussion on general topics and then about the patient's pet. The patient expressed that she missed her pets which were six cats at her home. She said PARO reminded her of them. She voluntarily talked to PARO by saying 'I love you PARO'. And the end of the session she mentioned that she would like to see PARO again. The patient was observed to be more comfortable with PARO on the second day. On the third session, the topic introduced by the psychologist was about her weekend activities, visits from family members, her plans after the rehabilitation program is complete and about return-to-work. She mentioned that she felt calmer. She wanted to get better so she can take care of her sick father. She expressed that she missed PARO after the second session and that it would be good to have PARO to accompany her before sleep. It was also observed that the patient showed more verbal communications with PARO.

4.1   Results of Screening Evaluations

Results on evaluations using the HAM-D, HAM-A, C-SSRS and PSQI instruments are tabulated in Table 1. For HAM-D, the scoring instructions are 0–7 = Normal, 8–13 = Mild Depression, 14–18 = Moderate Depression, 19–22 = Severe Depression

Table 1. Results from screening instruments

Screening types	Evaluation results	
	Pre-experimental	Post-experimental
Hamilton Depression (HAM-D)	13	8
Hamilton Anxiety (HAM-A)	17	6
Columbia Suicide (C-SSRS)	Denied	Denied
Quality of Sleep (PSQI)	Mild disturbance	Reduced disturbance

and ≥ 23 = Very Severe Depression. For HAM-A, a total score of <17 = Mild Severity, 18–24 = Mild to Moderate Severity and 25–30 = Moderate to Severe. For HAM-D, score of 13 in pre-experimental falls in the high end of mild depression category. The screening evaluation suggests mild insomnia (difficulty to sleep), mild depressed mood in terms of sadness, anxiety and pessimism about the future. For post-experimental, score reduced to 9 (lowest end of mild depression) with reduced anxiety, worrying and fear of future. In terms of depressed mood: tendency of crying had reduced too. Patient's sleep had improved where difficulty of falling asleep had reduced in post experiment. Patient reported that she fell asleep in less than 30 min. For HAM-A, the scores were 17 (pre) and 6 (post). Both fall under the mild severity anxiety. This is consistent with HAM-D (pre) results where patient reported moderate to severe anxious mood, tension, fear and concentration difficulty. In post-experiment HAM-D reported mild anxious mood, tension and mild depressed mood. Overall behaviour observation using HAM-A showed that in the pre-experimental, patient reported restlessness and agitation but in the post-experimental, patient only exhibited mild restlessness and agitation.

The C-SSRS instrument was used to further support HAM-D in the suicide section. In the pre and post screening, C-SSRS results support the scores using HAM-D where the patient denied suicide. This reflects that she had positive insight to recover and become better. C-SSRS scores also support findings from the image processing tool that detects the frequency of her smile. This will be covered in the next section. PSQI instrument that measures the quality of sleep also reflected congruency with results in HAM-D where evaluation result improved from mild disturbance before the experiment to reduced disturbance after the experiment. The level of depression and anxiety had reduced. Hence, the quality of sleep had improved after interaction with PARO. Even though the duration of three days was short, positive changes were observed.

4.2 Results of Video Processing

Sample of output results from the OpenCV smile detection mechanism is shown in Fig. 6. The graph in Fig. 6 differentiates the frequency of smiles in the 10-min therapy duration without PARO followed by the last 10-min therapy with PARO. This duration was chosen as this was when the patient has familiarized herself and felt comfortable with PARO. The results show that there is an increase in number of smiles with and without PARO in the first, second and third session. Findings from the image processing tool is consistent with results from C-SSRS scores where the patient still relate

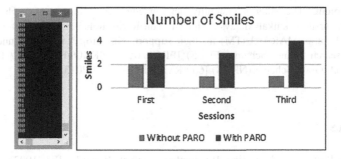

Fig. 6. Example of OpenCV output (left) and bar graph comparing the number of smiles (right)

to reality and has good judgment. Thus her smiles were natural and within normal limits. Patients with depression rarely smile. Thus, even though the total number of smiles were small in the pre and post-experiment, the finding is significant. The frequency of smile was the same in the first two sessions and increased in the third session suggesting that the patient felt calmer when interacting with PARO.

4.3   Overall Observation

The screening results suggest that PARO appeared to help the patient to manage her psychological distress. In the pre and post period, there was no other psychological intervention administered to the patient besides PARO to minimize the co-founding factors. The interactions with PARO show that patient could express her emotion and also personal issue easier during therapy with PARO. This might explain why the symptoms of anxiety and depression reduced. PARO acted as a platform to enable the patient to express herself. Though the number of smiles detected was few (maximum of 4 in the final session), it is considered as positive outcome for a patient with depression. The patient had enjoyed the companionship of PARO robot over the three days of exposure.

5   Conclusion

This study suggests that PARO effectively uplifts mood and helps the patient to be calm. PARO reacted to stroking by the patient by continuously blinking its eyes and moving its flipper. This continuous trigger prompted the patient to give respond to PARO throughout the session. Even though the patient was prescribed with anti-depressant, it was only meant to improve her depression symptoms and will not trigger her to smile. So, the increase in the number of smiles is a natural respond and within the normal limits of a typical person. For future works, studies with a larger sample of patients are needed as the effect of repetitive interaction with PARO needs to be investigated. Depression is a serious mental illness but often under-recognized [9]. Embedding state-of-the-art robotics technology into depression therapy shall give new hope to people affected with depression all over the world.

Acknowledgment. The authors gratefully acknowledge the Ministry of Higher Education Malaysia, Universiti Teknikal Malaysia Melaka, Tun Razak Rehabilitation Center Melaka and Universiti Teknologi MARA (UiTM) for their support. This project is funded under the Fundamental Research Grants Scheme (FRGS) [FRGS/1/2016/SKK06/FKP-AMC/ F00321] and Niche Research Grant Scheme (NRGS) [600-RMI/NRGS 5/3 (1/2013)].

References

1. Tiller, J.: Post-stroke depression. Psychopharmacology **106**, 130–133 (1992)
2. Gainotti, G., Antonucci, G., Marra, C., Paolucci, S.: Relation between depression after stroke, antidepressant therapy, and functional recovery. J. Neurol. Neurosurg. Psychiatry **71**, 258–261 (2001)
3. Macauley, B.L.: Animal-assisted therapy for persons with aphasia: a pilot study. J. Rehabil. Res. Dev. **43**, 357 (2006)
4. Marti, P., Bacigalupo, M., Giusti, L., Mennecozzi, C., Shibata, T.: Socially assistive robotics in the treatment of behavioural and psychological symptoms of dementia. In: The First IEEE/RAS-EMBS International Conference on Biomedical Robotics and Biomechatronics BioRob 2006, pp. 483–488 (2006)
5. Shibata, T., Coughlin, J.F.: Trends of robot therapy with neurological therapeutic seal robot. PARO. J. Robot. Mechatron. **26**, 418–425 (2014)
6. Clarke, D.J., Forster, A.: Improving post-stroke recovery: the role of the multidisciplinary health care team. J. Multid. Healthc. **8**, 433–442 (2015)
7. Shamsuddin, S., Abdul Malik, N., Hashim, H., Yussof, H., Hanapiah, F.A., Mohamed, S.: Robots as adjunct therapy: reflections and suggestions in rehabilitation for people with cognitive impairments. In: Omar, K., et al. (eds.) FIRA 2013. CCIS, vol. 376, pp. 390–404. Springer, Heidelberg (2013). doi:10.1007/978-3-642-40409-2_33
8. Wada, K., Ikeda, Y., Inoue, K., Uehara, R.: Development and preliminary evaluation of a caregiver's manual for robot therapy using the therapeutic seal robot Paro. In: 2010 IEEE RO-MAN, pp. 533–538 (2010)
9. Hum, L.C.: Management of Major Depressive Disorder. Clinical Practice Guidelines (2007)
10. Montano, C.B.: Recognition and treatment of depression in a primary care setting. J. Clin. Psychiatry (1994)
11. Aben, I., Verhey, F., Lousberg, R., Lodder, J., Honig, A.: Validity of the beck depression inventory, hospital anxiety and depression scale, scl-90, and hamilton depression rating scale as screening instruments for depression in stroke patients. Psychosomatics **43**, 386–393 (2002)
12. Posner, K., Brown, G.K., Stanley, B., Brent, D.A., Yershova, K.V., Oquendo, M.A., Currier, G.W., Melvin, G.A., Greenhill, L., Shen, S.: The Columbia-Suicide Severity Rating Scale: initial validity and internal consistency findings from three multisite studies with adolescents and adults. Am. J. Psychiatry **168**, 1266–1277 (2011)
13. Bradski, G., Kaehler, A.: Learning OpenCV: Computer vision with the OpenCV library. O'Reilly Media, Inc. (2008)
14. Freund, Y., Schapire, R.E.: A desicion-theoretic generalization of on-line learning and an application to boosting. In: Vitányi, P. (ed.) EuroCOLT 1995. LNCS, vol. 904, pp. 23–37. Springer, Heidelberg (1995). doi:10.1007/3-540-59119-2_166

Robot-Assistant Behaviour Analysis
for Robot-Child Interactions

Alina Zimina, Polina Zolotukhina, and Evgeny Shandarov[(✉)]

Tomsk State University of Control System and Radioelectronics
(TUSUR University), Tomsk, Russian Federation
evgenyshandarov@gmail.com

Abstract. The paper introduces research of interaction between a humanoid robot and a group of preschool children. As can be seen from the brief review the ultimate goal of any such researches is to create the most comfortable and native child-robot interaction with minimum involvement from the operator. But it is impossible to reach this goal without considering the peculiarities of robotics platforms as well as human psychology. A lot of number of features should be considered when conducting experiments and planning the scenario of interaction of a child with the robot.

Keywords: Humanoid robots · Child-robot interaction · Social robotics

1 Introduction

In cases where a child is in temporary isolation, for example, on clinical examination, it is difficult to provide a comfortable environment and plan time effectively, to include training sessions. Unfortunately, it is not always possible to attract people to this process, and it is not always effective. Most children do not feel comfortable in the presence of strangers. But all kids love toys. And the social robotics answers the questions of what should be the toy, and how it should interact with the child.

There are many definitions of a social robot:

- Socially evocative. Robots that rely on the human tendency to anthropomorphize and capitalize on feelings evoked, when humans nurture, care or involve with their "creation" [3, 4].
- Socially situated. Robots that are surrounded by a social environment which they perceive and react to. Socially situated robots are able to distinguish between other social agents and various objects in the environment [10].
- Sociable. Robots that proactively engage with humans in order to satisfy internal social aims (drives, emotions, etc.). These robots require deep models of social cognition [3, 4].
- Socially intelligent. Robots that show aspects of human-style social intelligence, based on possibly deep models of human cognition and social competence [5].
- Fong et al. [10] propose the term "socially interactive robot", which they define as a robot for which social interaction plays a key role in peer-to-peer HRI, different from other robots that involve "conventional" HRI, such as those used in teleoperation scenarios.

© Springer International Publishing AG 2017
A. Ronzhin et al. (Eds.): ICR 2017, LNAI 10459, pp. 219–228, 2017.
DOI: 10.1007/978-3-319-66471-2_24

From the definitions it is clear that a social robot should have the following characteristics: be able to perceive and express emotion, recognize themselves like robots and people to maintain a dialogue verbally and non-verbally, to maintain long-term relationships, possess personal traits, personality, ability to study and learn new behaviors. These requirements are setting the following challenges to this social robotics:

- Clarification of human social behavior models, in particular, affective (emotional) and cognitive components of behavior, learning and adaptation in social behavior, the development of intelligence, etc.
- Social interaction between humans and robots (research, forecasts, the development of ethical standards, etc.).
- Social interaction between robots (collaboration, mutual aid).

2 Background

2.1 Survey of Child – Robot Interaction Projects

Human – robot interaction (HRI) is an interdisciplinary area: robotics, engineering, computer science, psychology, linguistics, ethology and other disciplines investigating social behaviour, communication and intelligence in natural and artificial systems. Different from traditional engineering and robotics, interaction with people is a defining core ingredient of HRI. Such interaction can comprise verbal and/or non-verbal interactions [11].

Dautenhan [6] proposed that approaches to social interactions with robots can be categorized into three groups: robot-centred HRI, human-centred HRI and robot cognition-centred HRI. Robot-centred HRI means that a robot is an independent entity with its own goals and desires, and human interaction is only a means to achieve them. Human-centred HRI is primarily concerned with how a robot can fulfill its task specification in a manner that is acceptable and comfortable to humans. Robot cognition-centred HRI emphasizes the robot as an intelligent system, i.e. a machine that makes decisions on its own and solves problems it faces as part of the tasks it needs to perform in a particular application domain.

Child-robot interaction is mainly based on Human-centred HRI and Robot cognition-centred HRI. An example of CRI is the number of projects designed to investigate the interaction of children with robots: Aurora project and ALIZ-E project. The first project studies the use of robots for therapy and education of children with autism. According Dautenhan [6], several experiments were supplied with different platforms and, consequently, different types of child-robot interaction. The researchers conclude that, in the case of children with autism robot-human relationships as effectively as human-human. The second project is studied for a long interaction with children who undergo clinical examination. The paper by Belpaeme et al. [2] describes the adaptive social models and their implementation with the robot Aldebaran Robotics Nao. For example, the robot remembers the children with whom it spoke, their faces, names, age and other information to further the impression that the robot is not just a toy, but something that has consciousness and memory. When communicating, the robot expresses emotions through poses and sounds, the robot's vocabulary is understandable to children. Most of the children engage is playing with Nao as well as playing with other children, and, in addition, all children like to chat with the robot, and many of them responded to the robot truly personal questions. The same conclusions are made by Baxter et al. [1].

Another CRI experiment is described in the article [14]: 3–6 years old children learn the English words with the robot and without it. Three forms of learning are used when working with the robot: direct teaching, gesturing, and verbal teaching. The direct teaching study means that the word (usually a verb), that should be studied, is exemplified by the showing a movement, that impersonates the word, to the robot. The gesturing means that child shows the robot motion which corresponds to the word, and the verbal teaching is process when the child give a voice command to the robot and then the robot does the movement. Aldebaran Robotics Nao was also selected for interaction. In beginning it is revealed the optimal number of children in the group (10–15 people). The experiment shows that learning with the robot more is 30% effective and the most interesting form of learning is direct teaching.

Ros Espinosa et al. [13] describes other robot-child interaction scenarios: copy-dancing, "Simon says" (a player has to follow Simon's instructions), question-answer when the robot asks a question and waits for a response from the child. All of them are the basis of the development of playful designs for a robot that can take on different roles (such as instructor, companion, and playmate). The experiments were implemented on Aldebaran Robotics Nao.

Based on these projects it can be concluded that humanoid medium-height (up to 60 cm) robots interact with children most efficiently. Children like the robot and usually see them as toys or equals. Non-verbal communication was important: the more dynamic were the robot's movements, the easier was understanding and memorizing. The experimenter or observer must be present (whether explicit or implicit, via telepresence) during the robot-child interaction. But several speech-related questions remain unanswered. What should the voice and speech rate be like? Could the robot deliver a lecture?

2.2 Aldebaran Robotics Nao

As a research platform Aldebaran Robotics Nao used by more than 550 high-profile universities and research labs around the world. It is a small humanoid robot, measuring 58 cm in height, weighing 4.3 kg and having 25 degrees of freedom. It has a lot of sensors and actuators: 2 loudspeakers, 4 microphones, 2 cameras, a gyroscope, an accelerometer, and range sensors (2 IR and 2 sonar). The robot has an Intel Atom CPU core and connects externally via IEEE 802.11g WiFi or Ethernet. The Nao has a generally friendly and non-threatening appearance, which is therefore particularly well suited for child-robot interaction [12]. Nao SDK includes a voice synthesizer and a voice recognition system. Visual IDE Choregraphe is used for develop software applications.

3 Methods of Experiments and Interviews of Preschool Children

Preschool children have a number of psychological and behavioral characteristics that are necessary in order to obtain reliable results in the process of psycho-diagnostic testing. These features, in particular, include a relatively low level of consciousness and self-awareness. Most preschoolers' cognitive processes (such as attention, memory, perception, imagination and thinking) are at a relatively low level of development.

Preschoolers are not fully aware of their own personality and cannot properly evaluate their behavior. Staring at 4–6 years of age, children can evaluate themselves as a person, but still to a limited extent. So it is recommended to access the method of external peer review using experts as adults who know the child [8, 9].

Also preschoolers are not quite developed enough to complete personality questionnaires that contain direct self-assessment statements [9]. In general, the use of such questionnaires at pre-school age for psycho-diagnostic purposes should be kept to a minimum, and using them is unavoidable, every question should be explained in detail in a way that is understandable for a child.

Finally, attention should be paid to the characteristics of involuntary cognitive processes, for example, the volatility of involuntary attention, and the easy fatigability of children of this age. Therefore, tests should not be too long or requiring a lot of time. The best results can be obtained by observing children during the main activity at that age – during play [9].

Also, to obtain reliable results it is necessary to establish friendly contact and good understanding between the child and the experimenter. All testing methods developed for preschool children must be brought individually or in small groups of kindergarten children with teamwork experience. Typically, tests for preschoolers are presented orally or in the form of tests. Sometimes assignments may involve using a pencil and paper (for simple tasks) [16].

Let us consider the application of different methods children study, such as observation, quiz or interview, experiment and testing.

Observation is the one of the most important methods for children work. Many of the techniques commonly used in adult studies (test, experiment, survey) have limited use in children studies because of their complexity.

Observation has a lot of different options, which together provide a good variety of reliable information about children. Before monitoring what and how children do, one should identify the goal of observation and answer the questions: what is being done and for what, and what the results should ultimately be. Then the observation program needs to be developed and a plan needs to be designed to ensure that it leads researchers to the desired goal.

On the one hand, monitoring children is easier than adults, since children under the supervision usually act more natural and no not play special social roles inherent in adults. On the other hand, children, especially preschoolers, have an increased responsiveness and unstable attention and often switch between activities.

The method of **conversation** (or questions) can be used in working with children aged 4 years old or older when they already are sufficiently proficient with language, but in a very limited range. Pre-school children are not yet able to express their thoughts and feelings with words, so their answers are usually brief, formal and reproduce adult speech [15]. Difficulties may be caused by the fact that children do not always fully understand the questions addressed to them. Questions should be clear and interesting to the child, and in no case should give any clues. Conversation can only be used as an auxiliary, secondary method.

According to Drujinin [7], in children research, **experiment** is often one of the most reliable methods of obtaining reliable information about the psychology and behavior of the child, especially when monitoring is difficult and survey results may be questionable. Inclusion of children in the experimental situation of the game allows one to direct the child's response to acting on the basis of these reactions to judge whether a child hides from seeing or not being able to verbalize in the survey. The experiment in working with children can get the best results when it is organized and carried out in the form of games or familiar to the child classes – drawing, construction, guessing riddles, etc.

A **test** is a system of specially selected tasks that are given to children under strictly defined conditions. Each task is awarded with points. Assessment should be objective and not depend on the personal relationships of the experimenter. An important feature of practical application of tests that should be taken into account in conclusions, is getting used to the tests, and the relative variability, inconsistency of their results. Many children completing tests quickly get tired of repetitive and monotonous work, they unwittingly begin to change their answers or test procedure to get rid of satiety [16].

4 Interaction Scenarios and Experiments

The aim of the study was to determine the optimal parameters of child-robot interaction scenarios, the duration of stage cues between actions and so on. The result of the study was to be a modification of the script for the best possible interaction between a child and a robot [17, 18].

For the target audience, it was decided to select children aged 3 to 6. Children at this age tend to already know how to count using fingers, a part of the alphabet and know enough to act freely with adults. Because children feel more relaxed in the group we decided to interact with a group of children, rather than one child. The group included 5 children aged 4 to 5 years old. All children in the group were girls.

The chosen model focused on teaching mental arithmetic to children. The interaction scenario was developed as shown in Fig. 1.

A brief introductory lecture was delivered in order to establish closer contact with children and introduce them to the topic. The first part of the message was read out by one of the authors of this work, having experience with children as an animation actor. The second part of the presentation was doing the robot itself through its voice capabilities. In addition, the robot during a presentation actively gesticulated. It is assumed that this should help to establish closer contact between the robot and children.

The main part of the experiment actually began after a brief presentation. At this stage participation of adults was minimized. Since the NAO robot has only two

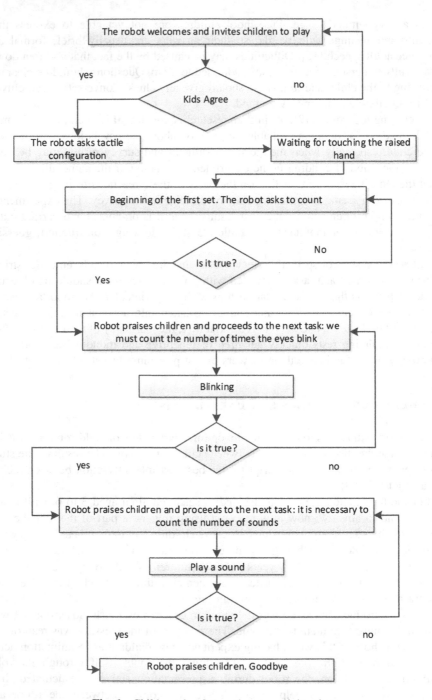

Fig. 1. Child – robot interaction scenario's scheme

channels of interaction allowing feedback: voice functions (synthesis and recognition), and tactile sensors (hands, feet, head), it was decided to include both types of interaction in the script. Thus, to establish robot – child contact the next episode was designed in the scenario. The child should shake robot's hand so that the tactile sensor is touched and NAO knows that someone is here.

As can be seen from the scenario of interaction, communication consisted of three parts. During the first part, the robot asked the children to count its fingers on both of its hands. The answer has to be spoken. The second part was based on the fact that the robot can blink, and the children have to count the number of these signals and answer the question "How many times have I blinked?" Finally, in the third part of the game children were asked to count the number of sounds. The role of the adult at this stage was limited only to help children in the performance of tasks, such as a hinting that it is easier to count using fingers.

It should be noted that the children successfully completed all the robot's tasks. The tasks were only difficult to complete when the answer exceeded the count of 7. Apparently, it is easier to count using only the fingers of one hand at that age.

After the main part of the experiment, children were given questionnaires to fill out, containing three questions:

- Did you like the robot?
- Who was a more interesting speaker?
- Did the robot tell the story clearly?

Of course, filling out questionnaires is not a very easy task at that age. Therefore, questionnaires also involved colorful pictures. View profiles shown in Fig. 2.

When children fill out the questionnaires adults actively help them by explaining the questions. At the end of the action, the robot thanked the children for participating in the experiment and performed a short dance. The children applauded to the robot's performance.

1. Do you like the robot?

2. Who was more interesting speaker?

3. Did the robot tell the story clearly?

Fig. 2. Questionnaire for children

Adults who were present at the event completed the questionnaire containing the following questions:

- Do you think the robot is a good addition to the educational process of pre-school children?
- How attentively do you think the children listened to the robot?
- Is the robot interesting for children?
- What do you think are the requirements to the robot that should be considered in robot-human interaction?

The entire experiment was video recorded and photographed.

5   Experiment Analysis

The material results derived from the experiment were video and photo materials and the completed questionnaires of children and adults.

The questionnaires of the 3 observing adults were analyzed. According to their questionnaires, the robot was interesting to children. It holds the child's attention longer (and the same can be noted in the video: the children focused their attention on the robot for longer periods than on the assistant.) A child perceives the robot as an equal in age, in contract to the teacher who has a bigger age difference with the child.

The observers pointed out that, first, the voice of the robot has to be softer and more human-like, so the children will better understand the robot. Second, in dealing with a child the robot has to talk using age-appropriate language, avoiding formal phrases and limiting the vocabulary to domestic and familiar words. Finally, tactile interaction is very important with a robot, it is necessary to introduce elements of interactivity and more robot movements during the dialogue (facial expressions, gestures).

In addition, the following problems were noted:

- Speech Recognition. Children do not pronounce the words clearly, and the robot did not always properly understand them.
- Short training scenario. As it turned out three math problems were insufficient for a full lesson.
- Position of the robot. During the experiment, the robot was standing on the table, so that the children perceived it on a par and it was difficult to reach when the robot said, "Give me five".

It may be added that adults relate to the robot with some caution and perceive it only as a supplement to the teacher in the educational process.

As adults actively helped the children in completing questionnaires, it has been suggested that the results of this survey will not be valid. On the one hand, adult help was absolutely necessary, because children can not read; on the other hand, adults can influence the answers given by children. In addition, as we have seen, the three-choice tests were not the best approach for a child. As it is easier to give one-word answers, all answers were identical:

- All children liked the robot.
- The robot talking about itself was more interesting than a man talking about the robot.
- The robot talked clearly to all children

Despite the non-relevant results of the survey, this part of the experiment was important: we were playing with children while making known the image and the functions of the robot.

Thus, the following conclusions were drawn from the analysis. In general, the interaction scenario was well chosen, because all children were actively involved in the process, gave the correct answers and showed no anxiety about having to communicate with robot. Initial is necessary, but it should be shorter, and its text should be made more "childish". It is possible and desirable to have the robot interact with the audience, for example by means of tactile sensors. The number of tasks that the robot gives to children seems optimal: the children do not look tired, and take a keen interest. The feedback channel for interaction should be made free of the drawbacks of voice control (low noise resistance, inaccurate recognition, long response times), perhaps it would be advisable to use the technical vision of the robot. Instead of filling out questionnaires after presentation, it is advisable to hand out informational materials to children, such as posters with the robot in it and its mnemonic function mappings.

6 Conclusions

This paper provides a survey of HRI research in the context of CRI. The results of experimental research of CRI was provided.

An experiment was carried out consisting of two parts. The first was passive listening and the second was an active game. The experiment analysis were defined new challenges to be investigated in the future. First we need to find new ways of child-robot interaction instead of voice input. Next we should develop of additional tests and experiments that help to compose adequate statistical picture of the developed scenarios.

References

1. Baxter, P., Belpaeme, T., Canamero, L., Cosi, P., Demiris, Y., Enescu, V.: Long-term human-robot interaction with young users. In: ACM/IEEE Human-Robot Interaction 2011 Conference (Robots with Children Workshop) (2011)
2. Belpaeme, T., Baxter, P., et al.: Multimodal child-robot interaction: building social bonds. J. Hum.-Robot Interact. 1(2), 33–53 (2012)
3. Breazeal, C.: Designing Sociable Robots. MIT Press, Cambridge (2002)
4. Breazeal, C.: Towards sociable robots. Robot. Auton. 42, 167–175 (2003)
5. Dautenhahn, K.: The art of designing socially intelligent agents-science, fiction, and the human in the loop. Appl. Artif. Intell. 12, 573–617 (1998)
6. Dautenhahn, K.: Socially intelligent robots: dimensions of human–robot interaction. Phil. Trans. R. Soc. b 362(1480), 679–704 (2007)

7. Drujinin, V.: Experimental Psychology. SPB, Saint-Petersburg (2002). (in Russian)
8. Elkonin, D.: Child Psychology. Practical Psycology Institute, Voronej (2011). (in Russian)
9. Elkonin, D.: Mental Development in Childhood. Prosveshenie, Moscow (1995). (in Russian)
10. Fong, T., Nourbakhsh, I., Dautenhahn, K.: A survey of socially interactive robots. Robot. Auton. Syst. **42**, 143–166 (2003)
11. Klingspor, V., Demiris, Y., Kaiser, M.: Human-robot-communication and machine learning. Appl. Artif. Intell. **11**, 719–746 (1997)
12. Nalin, M., Bergamini, L., Giusti, A., Baroni, I., Sanna, A.: Children's perception of a robotic companion in a mildly constrained setting: how children within age 8–11 perceive a robotic companion. In: Proceedings of the Children and Robots Workshop at the IEEE/ACM International Conference on Human-Robot Interaction (HRI 2011). IEEE, Lausanne, Switserland (2011)
13. Ros Espinosa, R., Nalin, M., Wood, R., Baxter, P., Looije, R., Demiris, Y., Belpaeme, T.: Child-robot interaction in the wild: advice to the aspiring experimenter. In: Proceedings of the ACM International Conference on Multi-modal Interaction, pp. 335–342. ACM, Valencia, Spain (2011)
14. Tanaka, F., Matsuzoe, S.: Children teach a care-receiving robot to promote their learning: field experiments in a classroom for vocabulary learning. J. Hum.-Robot Interact. **1**, 78–95 (2012)
15. Vallon, A.: Mental Development of the Child. Prosveshenie, Moscow (1967). (in Russian)
16. Vigotsky, L.: Pedagogical Psychology. Prosveshenie, Moscow (1991). (in Russian)
17. Zimina, A., Rimer, D., Sokolova, E., Shandarova, O., Shandarov, E.: The humanoid robot assistant for a preschool children. In: Ronzhin, A., Rigoll, G., Meshcheryakov, R. (eds.) ICR 2016. LNCS, vol. 9812, pp. 219–224. Springer, Cham (2016). doi:10.1007/978-3-319-43955-6_26
18. Gomilko, S., Zimina, A., Shandarov, E.: Attention training game with aldebaran robotics NAO and brain-computer interface. In: Ronzhin, A., Rigoll, G., Meshcheryakov, R. (eds.) ICR 2016. LNCS, vol. 9812, pp. 27–31. Springer, Cham (2016). doi:10.1007/978-3-319-43955-6_4

Context-Based Coalition Creation in Human-Robot Systems: Approach and Case Study

Alexander Smirnov[1,2], Alexey Kashevnik[1,2(✉)], Mikhail Petrov[1,2], and Vladimir Parfenov[2]

[1] SPIIRAS, 39, 14th Line, St. Petersburg 199178, Russia
{smir, alexey}@iias.spb.su, dragon294@mail.ru
[2] ITMO University, 49 Kronverksky Pr, St. Petersburg 197101, Russia
parfenov@mail.ifmo.ru

Abstract. The paper presents a context-based approach to coalition creation in human-robot cyber-physical systems. Cyber-physical systems tightly integrate physical and information spaces based on interactions between these spaces in real time. Mobile robots and humans are exchange information with each other in information space while their physical interaction occurs in physical space. The information space is organized based on Smart-M3 platform. This platform allows to organize ontology-based information and knowledge sharing for various participants based on publication subscription mechanism. For semantic interoperability support the ontology is used for problem domain modelling. The ontology formally represents knowledge as a set of concepts within a domain, using a shared vocabulary to denote the types, properties, and interrelationships of those concepts. For the implementation the point exploring and obstacles overcoming scenario has been chosen and implemented. This scenario covers two base cases for coalition creation: robot-robot and robot-human. Mobile robots have been constructed based on Lego Mindstorms EV3 Kit.

Keywords: Collaborative robotics · Coalitions · Human-robot interaction · Interoperability · Ontologies · Context management

1 Introduction

Nowadays, mobile robots become more and more popular in many areas. They are actively used for different tasks such as scouting, technological accidents and catastrophes, counterterrorism operations and patrolling [1]. The utilization of robots is useful in case of the impossibility of employment of human resources, in the presence of threats to health and human life, as well as to perform consuming tasks in adverse conditions. Sometimes, performing a task a robot tends to face obstacles on its path because of both heterogeneity and complexity of a landscape. Depending on the complexity of task it can be solved by a robot or human-robot coalition has to be formed for joint task solving. For example, in case of presence obstacles the robot can try to overcome it and in case of fail can ask human operator to take control and overcome the obstacle. The implementation of these options requires an appropriate

© Springer International Publishing AG 2017
A. Ronzhin et al. (Eds.): ICR 2017, LNAI 10459, pp. 229–238, 2017.
DOI: 10.1007/978-3-319-66471-2_25

approach to problem solving and interaction of robots with humans taking into account current situation (e.g., obstacle parameters).

This paper proposes context-based approach to coalition creation in human-robot cyber-physical systems. Context is defined as any information that can be used to characterize the situation of an entity. An entity is a person, place or object that is considered relevant to the interaction between a user and an application, including the user and application themselves [2]. For the case study the scenario of a point exploring and obstacles overcoming is proposed and implemented. Based on this scenario the coalition of robots and a human expert is formed for task solving. Each robot consists of several blocks with pairs of wheels and is equipped with ultrasonic sensors. Robots are able to detect obstacles, and overcome them (the height of the obstacle can be more than radius of the wheel). The interaction between robots and humans is implemented in Smart-M3 information sharing platform. This technology allows to organize ontology-based information and knowledge sharing for various participants based on publication subscription mechanism. The paper extend the previous authors work related to Smart M3-based robot interaction scenario for coalition work [3].

The rest of the paper is organized as follows. Section 2 describes related work in the area of mobile robots interaction. Section 3 presents context-based coalition creation approach. Section 4 propose a case study. Finally, Conclusion summarizes the paper.

2 Related Work

There are several research papers in the area of human-robot and robot-robot interaction for joint task solving have been considered and analyzed.

Paper [4] deals with an intelligent system of robot and human interaction. This system uses adaptive learning mechanisms to account for the behavior and preferences of a human. It includes algorithms for speech and face recognition, natural language understanding, and other ones to the robot could recognize human actions and commands. A robot used in this paper (Pioneer 3-AT) is equipped with multiple sensors, scanners and a camera. The paper focuses on the process of teaching the robot and its mechanisms of interaction with a human.

Paper [5] describes a coordinated process for the agents (in particular, mobile robots) jointly perform a task, using the knowledge about the environment, their abilities and possibilities for communication with other agents. The effective use of knowledge about the agents capabilities is implemented by using suitability rates which allow agent to choose the most appropriate action. This approach was tested during a football match between mobile robots. The interaction in this article is considered in terms of the choice of an agent to perform an action.

Paper [6] proposes a methodology for coordination of autonomous mobile robots in jams in congested systems. The methodology is based on reducing a robot's speed in jams and congested areas, as well as in front of them. Thus, the robots slow down detecting other robots in front of themselves. The effectiveness of the methodology was tested in the simulation of congested system with a jam.

Paper [7] discusses combining wireless sensor networks and mobile robots. Authors suggest layered swarm framework for the interaction of decentralized self-organizing complex adaptive systems with mobile robots as their members. This framework consists of two layers (robots and wireless LAN) and communication channels within and between layers. During the experimental verification of framework mobile robots independently reached the destination.

Authors of the paper [8] is devoted to the problem of robots group control in non-deterministic, dynamically changing situations. Authors propose a method for solving formation task in a group of quadrotors. The method makes possible to ensure accurate compliance with distances between quadrotors in the formation, as well as featuring low computational complexity.

Paper [9] shows a system of robots interaction based on BML – Battle Management Language. This language is intended for expressing concise, unambiguous orders read by both humans and robots. The orders are transmitted from the system of multiple robots and processed for further distribution specific commands between them.

Paper [10, 11] describes designing human-robot interaction, improving human situational awareness on the basis of an agent-oriented approach. This approach is intended to apply at movement of mobile agents (autonomous robots and cosmonauts in a protective gear) on the Moon's surface in the controlled territorial area.

Paper [12] considers the problem of program agents' access to the smart space based on Smart-M3 platform. This paper suggests a model of concurrent sessions to work with multiple smart spaces. The model allows an agent to use multiple connections and control objects in a single session. Moreover, it provides local storage, divided between session connections. The storage is synchronized using subscriptions.

The approaches used in papers [4, 6] require a well-defined functionality of robots, so the set of undertaken tasks is restricted. In paper [5] the knowledge of the capabilities of robots is generated dynamically during the execution of tasks what allows to manage the robots effectively. Thus, no limitation to robots functionality and dynamic determination of the options available can provide the widest range of tasks to execute. Paper [7] shows the effectiveness of the indirect interaction of robots, that is, the interaction through an intermediary that is Smart-M3 platform in the paper. Paper [8] provides a formal model that can be used for interaction. Paper [9] shows the effectiveness of using specialized language for robots control. Papers [10, 11] demonstrates the applicability of the multi-agent systems approach for robots and humans interaction in various fields. Finally, paper [12] proposes a model of interaction between agents and a smart space.

3   Context-Based Coalition Creation

Coalition creation process requires semantic interoperability support between potential participants: mobile robots and humans. The process is based on cyber-physical systems concept. Cyber-physical systems tightly integrate physical and cyber spaces based on interactions between these spaces in real time. Mobile robots and humans are exchange information with each other in information space while their physical interaction occurs in physical space. It is needed to create the model of problem domain and support the interaction of potential participants based on this model. One of the possible approach to problem domain modelling is the ontological approach. The ontology formally represents knowledge as a set of concepts within a domain, using a shared vocabulary to denote the types, properties, and interrelationships of those concepts. For current situation modelling and reducing the search space in the coalition creation process the utilization of context management technology is proposed. The aim of this technology is a context model creation that is based on ontology of problem domain and information about current situation in physical space. Context is defined as any information that can be used to characterize the situation of an entity. An entity is a person, place or object that is considered relevant to the interaction between a user and an application, including the user and application themselves [2]. Context is suggested being modeled at two levels: abstract and operational. These levels are represented by abstract and operational contexts, respectively. The process of coalition creation based on abstract and operational contexts is shown in Fig. 1.

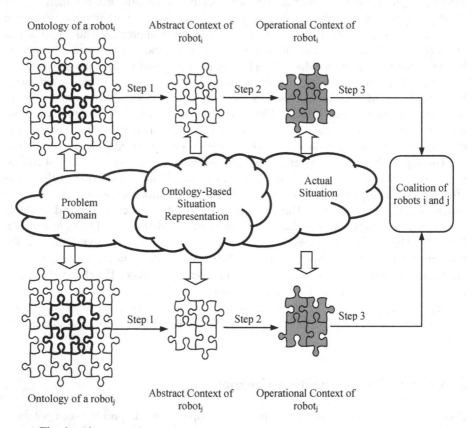

Fig. 1. Abstract and operational contexts for coalition creation of mobile robots

Abstract context is an ontology-based model of a potential coalition participant related to the task that formed coalition of robots have to be implement that is build based on integrating information and knowledge relevant to the current problem situation.

Operational context is an instantiation of the domain constituent of the abstract context with data provided by the contextual resources. Thereby the coalition creation is implemented in three steps. The first step includes the abstract context creation that covers the selection of knowledge relevant to the task from the potential coalition participant ontology. The second step includes the process of concretization of these knowledge by information accessible in information space for operational context formation. The operational context is published in information space and becomes accessible for other potential coalition participants. On the third step, the coalitions of mobile robots and humans are created based on their operational context intersections (Fig. 2).

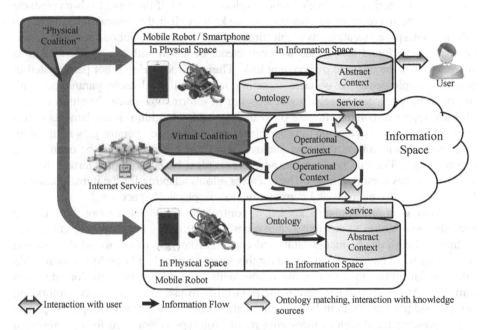

Fig. 2. Conceptual model for mobile robots and humans interaction for coalition formation

Information space is organized based on black board architecture that provides possibilities for potential coalition participants implement indirect interaction. Thereby, virtual coalition is created in information space and then physical coalition appears in physical space (mobile robots and humans implement the joint task). Interaction of potential coalition participants in information space is implemented based on ontology based publish/subscribe mechanism that provides possibilities for mobile robots and humans publishing their information and knowledge and subscribing on interesting information using ontologies. When a mobile robot or human registers to be a potential coalition participant it uploads the own ontology to the information space.

This ontology formalizes the main robot capabilities and constraints that have to be satisfied to use the capabilities. The human mobile device is used to represent him/her in information space, describes the human capabilities and tasks to be implemented. Mobile device also contains the human profile that provide to the information space the human preferences. Thereby, the interoperability for robots and humans interaction is supported based on open information space and ontology-based publish/subscribe mechanism. A potential coalition participant can participate in joint task if the ontology in information space is matched with own ontology. The ontology matching procedure is described in details in [13].

4   Case Study

For the case study the point exploring and obstacles overcoming scenario is proposed. During this scenario a six-wheeled robot explores a point. If the robot finds an obstacle that has to be overcoming to complete the task, it publish the context information to information space (location, available time). Then measuring robots get notification about obstacle found and decide who can create coalition with six-wheeled robot to solve together the obstacle overcoming task. Then the measuring robot participated in coalition implements the obstacle scanning, and publish obstacle parameters with information space. Six-wheeled robot searches in information space algorithm for this obstacle type overcoming. In case of availability this algorithm, it implements corresponding actions to overcome the obstacle. If not the human operator gets notification that he/she has to control the robot manually since the obstacle cannot be overcoming automatically. For this purposes an adaptive context-based control interface for the smartphone has been developed. The interface allows adapting for the human based on his/her role at the moment and current situation in physical space.

If more than one human is needed to control the six-wheeled robot the interface provides possibilities to distribute robot control functionality between different groups of humans, e.g., driver, manipulator, and expert. Prototypes of six-wheeled robot and measuring robot have been developed for this scenario based on Lego Mindstorms EV3 kit. This kit allows to design easily robots with required functionality for education purposes. At the same time, there is the possibility to use electronic units, motors and sensors and program them in Java. Figure 3 shows the six-wheeled robot prototype and Fig. 4 presents the design of measuring robot prototype constructed for the presented above scenario.

Fig. 3. Six-wheeled robot prototype

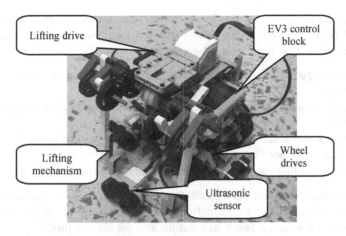

Fig. 4. Measuring robot prototype

Smart-M3 platform is used for information space implementation in the considered case study. It involves storing data in the form of RDF-ontologies, in which information is presented as a set of triples of "subject-predicate-object". A triples subject describes an entity such as an object, a device or a person. This entity has a property or can perform an action described in the triple. A triples object describes an entity, which the subject is associated with, that is its function or property. The object of one triple may be the subject of another one. A triples predicate describes the type of relationship between subject and object. It can determine the type of functions performed by the subject, or the correlation between subject and object.

A simple example of a triple stored in the ontology is <*"robot"*, *"task"*, *"goToLocation"*>. The subject *"robot"* describes a robot, which information in the triple relates to, the predicate *"task"* defines what this information is, and the object *"goToLocation"* is a particular *"task"* for the subject *"robot"*. At the same time *"goToLocation"* is also a subject in the triple <*"goToLocation"*, *"coordinates"*, *"40; 200"*>.

Devices connected to the information space based on Smart-M3 platform are able to "subscribe" to a triple to get notification when the needed information is published in the information space. In the case of a new triple satisfying pattern all devices, which are subscribed to triples of this pattern, are notified.

In order to overcome obstacles in accordance with the present scenario, the robot was constructed from several blocks. Due to the design the robot could climb an obstacle gradually. First, it lifted and fixed the front block, then basing on the front and back blocks it raised the middle one. All blocks were equipped with a pair of drive wheels; in addition, the middle block had a pair of wheels without a drive for balance. An ultrasonic sensor for measuring the distance to objects was also mounted on the central block. When the sensor detects obstacles in the front of the robot, automatic stop occurred.

Figure 5 shows the scenario for coalition creation for forward moving and encountered obstacles overcoming. The presented diagram shows five main components of the system that implements the scenario.

- Smartphone application publishes tasks to perform and shows the status of performance.
- Image processing service is designed to retrieve information about an object or area from a photo.
- Information space is an infrastructure to information sharing by mobile robots.
- The six-wheeled robot that overcome an obstacle.
- Measuring robot that measure the obstacle found by six-wheeled robot.

When starting each robot publishes information about its functions and limitations from its own ontology in the information space. Moreover, while executing tasks a robot publishes information about its location and the current task.

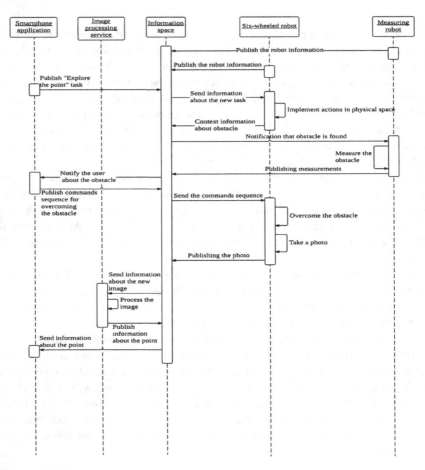

Fig. 5. The scenario of coalition creation for forward moving and encountered obstacles overcoming

For example, a user publishes task "Explore a point" in the information space. Since the robot is free, that is, it does not perform any other tasks, its interaction service determines whether the robot is to perform published task. For this purpose, the service follows the information about the task and about other robots. If a robot is the closest one to the point, it proceeds to the task.

When a robot detects an obstacle in the way, it automatically stops and publish in information space information about the obstacle. If there is no information about such an obstacle, the user takes control over the six-wheeled robot. As a result, the commands that are to be done in order to overcome this obstacle are consistently being published in the information space. After overcoming the obstacle the robot continues to perform the task and finds the second obstacle that is similar to the first one. Since the information about it is already in the obstacles templates database, the robot overcomes it without the user's help.

When the robot reaches the destination, and receives command "Take a photo", it takes a snapshot of terrain and sends it to the interaction service, which in its turn publishes an image in the information space. Then, the image processing service receives information about the new image and processes it. After receiving information about the point, the service publishes it in the information space, and then information is sent to the user.

5   Conclusion

The paper presents the context-based coalition creation approach in human-robot cyber-physical systems. The approach is based on ontology management, and context management technologies and supports semantic interoperability between mobile robots and humans for coalition creation. For the implementation the point exploring and obstacles overcoming scenario has been chosen and implemented. There are two mobile robots and a human ex pert is participated in the considered scenario. Mobile robots have been constructed based on Lego Mindstorms EV3 Kit. The scenario shows applicability of the proposed approach for coalition creation of human-robot coalition.

Acknowledgements. The presented results are part of the research carried out within the project funded by grants # 16-07-00462, 16-29-04349, 17-07-00247 of the Russian Foundation for Basic Research, program I.31 of the Russian Academy of Sciences. The work has been partially financially supported by Government of Russian Federation, Grant 074-U01.

References

1. Teja, S., Harsha, S., Siravuru, A., Shan, S., Krishna, K.: An improved compliant joint design of a modular robot for descending big obstacles. In: AIR 2015 Proceedings of the 2015 Conference on Advances in Robotics (2015)
2. Dey, A., Salber, D., Abowd, G.: A conceptual framework and a toolkit for supporting the rapid prototyping of context-aware applications. Hum.-Comput. Interact. **16**(2), 97–199 (2001)

3. Smirnov, A., Kashevnik, A., Mikhailov, S., Mironov, M., Petrov, M.: Smart M3-based robot interaction scenario for coalition work. In: Ronzhin, A., Rigoll, G., Meshcheryakov, R. (eds.) ICR 2016. LNCS, vol. 9812, pp. 199–207. Springer, Cham (2016). doi:10.1007/978-3-319-43955-6_24

4. Sekmen, A., Challa, P.: Assessment of adaptive human–robot interactions. Knowl.-Based Syst. **42**, 49–59 (2013)

5. Ibarra-Martínez, S., Castán-Rocha, J., Laria-Menchaca, J., Guzmán-Obando, J., Castán-Rocha, E.: Reaching high interactive levels with situated agents. Ingeniería Investigación y Tecnología **14**, 37–42 (2013)

6. Hoshino, S., Seki, H.: Multi-robot coordination for jams in congested systems. Knowl.-Based Syst. **61**, 808–820 (2013)

7. Li, W., Shen, W.: Swarm behavior control of mobile multi-robots with wireless sensor networks. J. Netw. Comput. Appl. **34**, 1398–1407 (2011)

8. Ivanov, D., Kapustyan, S., Kalyaev, I.: Method of spheres for solving 3D formation task in a group of quadrotors. In: Ronzhin, A., Rigoll, G., Meshcheryakov, R. (eds.) ICR 2016. LNCS, vol. 9812, pp. 124–132. Springer, Cham (2016). doi:10.1007/978-3-319-43955-6_16

9. Remmersmann, T., Tiderko, A., Schade, U., Schneider, F.: Interacting with multi-robot systems using BML. In: 18th International Command and Control Research and Technology Symposium, Alexandria, Virginia, USA (2013)

10. Karpov, A., Kryuchkov, B., Ronzhin, A., Usov, V.: Designing human-robot interaction in a united team of cosmonauts and autonomous mobile robots on the lunar surface. In: Proceedings 26th International Conference Extreme Robotics (ER-2016), St. Petersburg, Russia, pp. 71–75 (2016)

11. Kryuchkov, B., Karpov, A., Usov, V.: Promising approaches for the use of service robots in the domain of manned space exploration. SPIIRAS Proc. **32**(1), 125–151 (2014)

12. Lomov, A.: Ontology-based KP development for smart-M3 applications. In: Proceedings of the 13th Conference of Open Innovations Association FRUCT and 2nd Seminar on e-Tourism for Karelia and Oulu Region, Petrozavodsk, Russia, pp. 94–100 (2013)

13. Smirnov, A., Kashevnik, A., Shilov, N., Balandin, S., Oliver, I., Boldyrev, S.: On-the-fly ontology matching for smart M3-based smart spaces. In: Proceedings of the First International Conference on Mobile Ubiquitous Computing, Systems, Services and Technologies (UBICOMM 2010). Florence, Italy, 25–30, pp. 225–230 (2010)

Analysis of Overlapping Speech and Emotions for Interaction Quality Estimation

Anastasiia Spirina[1](✉), Olesia Vaskovskaia[2], and Maxim Sidorov[1]

[1] Institute of Communications Engineering, University of Ulm, Ulm, Germany
{anastasiia.spirina,maxim.sidorov}@uni-ulm.de
[2] Reshetnev Siberian State University of Science and Technology,
Krasnoyarsk, Russia
olesia.vaskovskaia@sibsau.ru

Abstract. Designing of indicators for problem detection during the dialogue is an important aspect for improving the systems user-adaptability in such rapidly expanding spheres as Spoken Dialogue Systems (SDSs) and Social Robotics (SR). Moreover, it is important not only for human-robot/computer spoken interaction (HCSI), but also for human-human conversation (HHC) as a way of service quality improvement in call centres. There are metrics, which may be used for both fields: HCSI and HHC. One of these metrics is customer/user satisfaction (CS), which modification is called Interaction Quality (IQ). Analysis of CS/IQ and human behaviour in some problematic situations (according to CS/IQ score) may be useful for further developing of an SDS that would be more human-like. Our research is focused on such essential parts of speech as emotions and overlapping speech. In this paper we analyse an impact of these speech features on automatic IQ estimation for HHC. Afterwards, we compare the obtained results with the result for HCSI.

Keywords: Human-human interaction · Task-oriented dialogues · Performances

1 Introduction

One of the aims of SDSs and SR developers is to make these systems more human-like. Ideally, these systems should be user-adaptive and more flexible in communication with users as real humans. Especially it is important for SR. Customer/user always was, is and will be the main indicator for further improvement/development direction of services or systems.

To reflect users opinion or to reveal a problem during dialogue, which almost always is the result of a mismatch of users expectations and the reality, there have been developed various metrics and indicators. One of such indicators is emotional state. It can be as an independent indicator and as an input value for calculation of other indicators. IQ model, which was designed for SDSs [1] an

A. Ronzhin et al. (Eds.): ICR 2017, LNAI 10459, pp. 239–249, 2017.
DOI: 10.1007/978-3-319-66471-2_26

adapted for HHC assessment [2], has as one of the input emotional state. Emotions may be negative as a result of a mismatch of user's (customer's) expectations and the reality. In the result of the further analysis, the specific reasons of such problem can be determined. But sometimes the negative emotions can confuse a research, due to such emotions could be caused by external factors. Moreover, some people can be overly emotional or not enough emotional. That is why, all these reasons can lead to the false view. According to [3] emotions may provide rich information for better prediction of further consumer's consumption behaviour. Another essential part of our speech is overlapping speech, which also may be used as an indicator of the dialogue quality.

Our work, described in this paper, was aimed to assess a benefit of using overlapping speech and emotions for IQ modelling for HHC and to find similarities or differences between IQ modelling for HCSI and HHC. We guess that such investigation can help then to improve IQ model both for HHC and for HCSI.

The paper is organised as follows. A concise observation of related work is given in Sect. 2. The utilized spoken corpus is shortly described in Sect. 3. Afterwards, information about the experimental setup is provided in Sect. 4. The following Sect. 5 presents the obtained results which are then discussed in Sect. 6. Finally, conclusions and future work are performed in Sect. 7.

2   Related Work

CS [4] is one of the widely used significant metrics, which helps to evaluate the service quality and SDS performance. Usually, it is estimated at the end of the calls handling the customer feedbacks via different interviews or questionnaires. In [5] the authors describe an attempt at automatically CS assessment using different call's features. The CS model is based on structured, prosodic, lexical and contextual features. It should be mentioned that lexical features include sentiment words, which may reflect the speaker's emotion or affect. According to the results obtained in [5] exclusion of this feature category leads to accuracy reduction in comparison with the results, which were obtained using all feature categories. However, we cannot make any conclusions about emotional state importance, because the lexical feature category includes more than only sentiment words. It should be pointed out that [5] provides the results of experiments of CS assessment in the middle and at the end of the calls.

Another attempt to evaluate automatically user satisfaction using emotions is presented in [6]. Authors define user satisfaction as "the final state of a sequence of emotional states: positive, negative, neutral".

In contrast to CS, Interaction Quality [1,7] allows to estimate dialogue performance at any point during the interaction. It is important to note that it was originally designed to control an SDS performance during ongoing spoken interaction as an analogue of CS. Later, it was adapted to HHC [2]. The IQ model for HCSI is based on the various features, including emotions. All experiments for IQ modelling for HCSI was performed on the LEGO corpus [8,9], which contains the following emotion set: angry, slightly angry, very angry, neutral, and friendly. All features, which are used for IQ modelling in [7], can be subdivided into the following groups: the automatic speech recognition parameters (ASR), features from the language understanding module (SLU), the dialogue manager-related parameters (DM), information about dialogue acts (DACT), emotional state, and user-specific information. The results, describing in [7], shows that the contribution of emotions to overall performance is not statistically significant according to the Wilcoxon Signed-Rank Test [10].

3   Corpus Description

The spoken corpus which were utilized for IQ modelling experiments includes 53 dialogues between four employees and fifty-three customers. Afterwards, all dialogues were split into 1,165 exchanges [11], which consist of agent's and customer's turn and possible overlaps. In turns, each turn/overlap is described by approx. 400 features, including acoustic features (384-dimensional feature vector, extracted by *OpenSMILE* [12], which was used for Interspeech 2009 Emotion Challenge [13]), speech duration, emotions, gender and other. Furthermore, there are features, which describe exchange itself: duration, number of overlaps, "who starts the exchange" and other. Moreover, the feature set comprises the window and dialogue parameter levels, which describe the last n exchanges and the complete dialogue up to the current exchange correspondingly. These levels contain some statistical information. In our study the window level covers the three last exchanges with respect to the current exchange.

3.1   Interaction Quality

All exchanges were accompanied with two IQ score labels, which are based on the different IQ-labeling guidelines. The rules for both annotation approaches can be found in [2].

The first approach *IQ1* is based on an absolute scale and is similar to the annotation guideline for HCSI [7]. Hence, this scale should include five classes (1-bad, 2-poor, 3-fair, 4-good, 5-excellent), but in the corpus only three classes are presented (with the IQ scores "3", "4", "5"), where the biggest class (the IQ score "5") covers 96.39% of all exchanges, the smallest class (the IQ score "3") includes only four observations.

In contrast to the first approach with the absolute scale, the second approach *IQ2* is based on a scale of changes, which is transformed into an absolute scale. The scale of changes consists of the following scores: "−2", "−1", "0", "1", "2", "1_abs" (the last score is in the absolute scale). Then, with the assumption that all dialogues start with the IQ score "5" in absolute scale (from the first approach), the obtained labels were transformed into an absolute scale. In our case there are four scores: "6", "5", "4", "3". Similar to the first approach the classes are also unbalanced with 88.24% of exchanges from the majority class "5". While the second biggest class "6" consists of 8.24% of all data. Concerning the smallest class "3", it contains four exchanges as the same class from the first approach [14].

3.2   Emotions

Each agent/customer speech fragment was manually labeled with three different emotion sets. These sets were chosen from [15] and adapted for our study. Thus, the first set (denote it *em1*) contains the following emotion categories: angry, sad, neutral, and happy. The following original set from [15] includes anxiety, anger, sadness, disgust, boredom, neutral, and happiness. But in our set *em2* there are all categories except anxiety. The third emotion set *em3* includes anger, sadness, disgust/irritation, neutral, surprise, and happiness. It should be mentioned that not all categories are presented in the corpus.

Afterwards, each original emotion set was subdivided into neutral and other emotions (denote them as *em{1,2,3}2*) and into negative, neutral, and positive emotions (denote them as *em{1,2,3}3*). The sets *em1* and *em13* are equals, because *em1* is presented in the corpus by such categories as: sad, neutral, and happy, which correspond to negative, neutral, and positive categories. From the destributions of the emotion labels depicted on Fig. 1, we can see that *em2* and *em3* differ from each other by the categories "boredom" and "surprise". As we can see: some speech fragments from *em2* categories "neutral" and "happiness" go to the categorie "surprise" from *em3*. The same situation is also with *em1* and *em2*, which differ from each other by categories "irritation" and "boredom".

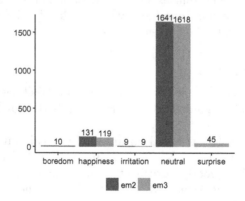

Fig. 1. Distribution of the emotion labels for *em2* and *em3*

4   Experimental Setup

In our work we have two approaches for IQ labelling and eight different emotion sets, that is why we have generated sixteen datasets. Besides, we have formed datasets without information about emotions for both IQ annotation approaches. Thus, the total number of different sets is eighteen for task of emotions role detection in IQ modelling for HHC. It means that for each set we used definite label set for emotions as input variables and definite IQ label's type as output.

The exchange's IQ score prediction task can be presented as a classification problem. For our experiments we have chosen the following classification algorithms: Kernel Naive Bayes classifier (NBK) [16], k-Nearest Neighbours algorithm [17] with the dimensionality reduction technique Principal Component Analysis [18] (kNN_PCA), L2 Regularised Logistic Regression (LR) [19], Support Vector Machines [20,21] trained by Sequential Minimal Optimization (SVM) [22], Multilayer Perceptron [23] (MLP), J48 algorithm (an open source Java implementation of the C4.5 algorithm) (J48) [24,25].

Partially, the choice of such algorithms as J48, NBK, LR, SVM can be justified by [5], where these algorithms were used for measuring CS. The following algorithm, namely (kNN_PCA), has shown encouraging results in [26]. For some algorithm we have utilized the default settings, which are implemented in *Rapidminer*[1] and *WEKA* [27]. For other we have optimised some parameters by the grid optimisation with F_1-score [28] maximization.

For the first four mentioned above algorithms we have used the same parameter settings, as in [26]. For MLP and J48 the settings can be found in [14].

Concerning the study of overlapping speech impact, we have performed all experiments with the first four aforementioned classification algorithms. Also we generate two sets for each emotion set. From one of them we have excluded all features for overlapping speech. While for the second set we have deleted only acoustic features for overlapping speech, keeping only information about duration of the fragments with overlaps and statistical information for window and dialogue levels.

To get a statistically reliable results we have perfromed 10-fold cross-validation. After splitting on training and testing sets we have applied one more inner 10-fold cross-validation on training set for the grid parameter optimization with F_1-score [28] maximization.

5   Results

To estimate classification performance of the applied methods we have used macro-average metrics: F_1-score, unweighted average recall (UAR) [29], and accuracy. It means, that the values of these metrics were averaged over ten computations on different train-test splits.

The obtained results are depicted for both *IQ1* and *IQ2* and all selected classification performance metrics on Figs. 2 and 3. The first aforementioned figure, namely Fig. 2, presents the box plots, which are based on the best achieved results for each dataset. Thus, the best achieved results in terms of F_1-score for *IQ1* (0.581) and *IQ2* (0.623) have been achieved with kNN_PCA on *em2* and *em33* respectively. Moreover, kNN_PCA has shown the best results for *IQ1* (0.606) on *em2* and *IQ2* (0.589) on *em3* in terms of UAR. Whereas, in terms of accuracy the best results for *IQ1* (0.976) on *em3* and *IQ2* (0.941) on *em2* have been achieved with SVM. In turn, the result for *IQ2* for all chosen performance

---

[1] http://rapidminer.com/.

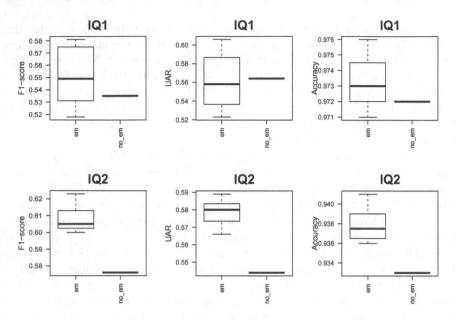

Fig. 2. The best achieved results for each dataset with and without emotions in terms of F_1-score, UAR, and accuracy

metrics have been obtained with SVM. Unlike *IQ2*, the results for *IQ1* in terms of F_1-score and UAR have been reached with kNN_PCA. The exception was only accuracy, where the best result has been obtained using NBK [14].

The second aforementioned figure contains the graphics, which reflect the achieved results on different datasets (eight sets with emotions and one set without emotions) for each classification method. The big black dots highlight the results, which were achieved on the dataset without emotion labels. It was done to understand whether emotion sets are beneficial or not depending on classification methods.

The results which have been gained for overlaps study can be found on Fig. 4. The results on the datasets with and without all features for overlapping speech were gained with kNN_PCA, while the results for the set without acoustic features for overlapping speech were reached on the LR-based IQ model.

6   Discussion

As we previously mentioned CS is usually measured at the end of the calls, while IQ score can be estimated at any point of ongoing interaction. But nevertheless, the IQ metric may be used for assessing the whole dialogue based on the achieved estimation on each exchange. The IQ score for the whole dialogue can be calculated using the following simple formula (1):

$$IQ_d = \frac{\sum_{i=1}^{n} IQ_i}{n} \tag{1}$$

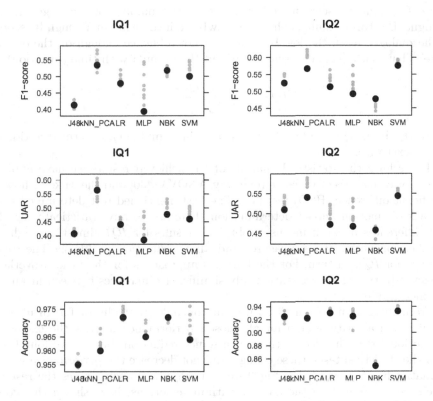

Fig. 3. The graphics of the obtained results for *IQ1* and *IQ2* in terms of F_1-score, UAR, and accuracy. The big black dots perform the results, which have been achieved on the dataset without emotion labels. Whereas the small grey dots mark the results, obtained with using information about emotions (eight emotion sets)

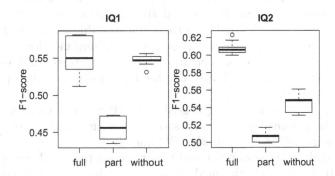

Fig. 4. The best achieved results for each dataset with all features (full), with the exception of the acoustic features (part), and with the exception all features (without) for overlaps in terms of F_1-score, UAR, and accuracy

where IQ_i is an IQ score for i^{th} exchange, n is a number of exchanges in the dialogue. But this formula (1) has the drawback: if there are many high IQ scores in the dialogue, then the low IQ score can not influence enough on the overall score. That is why it makes sense to use in the formula with time (duration):

$$IQ_{dt} = \frac{\sum_{i=1}^{n} IQ_i \times t_i}{\sum_{i=1}^{n} t_i} \tag{2}$$

where IQ_i is an IQ score for i^{th} exchange as in formula (1), t_i is the duration of the ith exchange.

To perform the statistical analysis of the achieved results we have applied the one-way analysis of variance (one-way ANOVA) [30] and the Tukey's honest significant difference (HSD) test [31]. The first mentioned test determined that for task of emotion impact determination there are no any statistically significant differences between means of obtained results for $IQ1$ almost through all classification performance measures and all classification problems. The same result is for $IQ2$. In turn, For the both IQ approaches for the task of overlaps impact detection there are statistically significant differences between means of obtained results.

From the graphics from Fig. 3 we can conclude, that almost for all utilized classification algorithms and for all chosen performance metrics the exclusion of emotional state has led to the performance reduction, but, as we found out from the statistical tests, these changes did not decrease the performance values dramatically. Moreover, from Fig. 2 we can see, that in some cases the results, which were obtained on the dataset without emotions, have shown the worst result among the all datasets. But, nevertheless, for $IQ1$ the dataset without emotions has shown the middle results in terms of different classification performance metrics.

However, from Fig. 3 we can see that for $IQ2$ in all the cases the results with emotions are constantly better than without, despite the fact that for $IQ1$ it is no so. We believe that this is due to the emergence of the new class "6" for $IQ2$. The objects were defined in this class, if the interaction was on neutral/good level (with the IQ score "5") and then the agent was able to win the favor of the customer. The agent can make a joke or say something nice for the customer. In this situation the IQ score and emotional state of customer and agent should be improved simultaneously. That is why for $IQ2$ we can see this dependence in comparison with $IQ1$.

Although, HHC is more complicated than HCSI and, thereby, they have differences, the obtained results have shown, that emotional state is not very important for modelling IQ for both HHC and HCSI.

But nonetheless, the obtained result is not enough to exclude emotional state from the further research in this field, because it can be only coincidence depending on corpora. To prove the state, that emotions may be excluded from the further IQ modelling process, an additional research is needed.

From Fig. 4 we can suggest that features, which describe overlaps duration, statistical information about overlaps on the window and the dialogue levels, may bring noise into the data, because after removing this features from corpus, the results increased. Also we can proposed that acoustic features of the fragments with overlapping speech may be useful for IQ modelling for HHC.

7   Conclusions and Future Work

In this paper we analysed the role of emotions and overlaps in IQ modelling for HHC. The impact of emotions is not significant, similar to the results described in [7]. We could not reveal any statistically significant difference between the results, which were obtained on the datasets with and without emotions. Partially it might be explained by the fact, that the used corpus is highly unbalanced and all labels were annotated only by one expert rater. That is why, further research of this question is necessary. Moreover, we could not find any tendency in results with different emotion sets. It means, that there is no a definite emotions set, on which all the best results have been achieved. Some results are the same to the results for HCSI described in [7]. However, we have revealed that the acoustic features of the fragments with overlapping speech may be useful for further IQ modelling for HHC, and as a result for HCSI and SR. Moreover, we have shown, that although the IQ metric was designed for measuring exchanges, it may be used for assessing the dialogue in general.

As a future direction we plan to extend experiments with the achieved results to get more details about IQ. Also, the techniques for unbalanced data should be performed.

Acknowledgments. The work presented in this paper was partially supported by the DAAD (German Academic Exchange Service), the Ministry of Education and Science of Russian Federation within project 28.697.2016/2.2, and the Transregional Collaborative Research Centre SFB/TRR 62 "Companion-Technology for Cognitive Technical Systems" which is funded by the German Research Foundation (DFG).

References

1. Schmitt, A., Schatz, B., Minker, W.: Modeling and predicting quality in spoken human-computer interaction. In: Proceedings of the SIGDIAL 2011 Conference, pp. 173–184. Association for Computational Linguistics (2011)
2. Spirina, A., Sidorov, M., Sergienko, R., Schmitt, A.: First experiments on interaction quality modelling for human-human conversation. In: Proceedings of the 13th International Conference on Informatics in Control, Automation and Robotics (ICINCO), vol. 2, pp. 374–380 (2016)
3. Wang, J.: From customer satisfaction to emotions: alternative framework to understand customer post-consumption behaviour. In: Proceedings of the 2012 International Joint Conference on Service Sciences, pp. 120–124 (2012)
4. Maar, B., Neely, A.: Managing and Measuring for Value: The Case of Call Centre Performance. Cranfield School of Management, UK (2004)
5. Park, Y., Gates, S.C.: Towards real-time measurement of customer satisfaction using automatically generated call transcripts. In: Proceedings of the 18th ACM Conference on Information and Knowledge Management, pp. 1387–1396 (2009)

6. Chowdhury, S.A., Stepanov, E.A., Riccardi, G.: Predicting user satisfaction from turn-talking in spoken conversations. In: Proceedings of INTERSPEECH 2016, pp. 2910–2914 (2016)
7. Schmitt, A., Ultes, S.: Interaction quality: assessing the quality of ongoing spoken dialog interaction by experts and how it relates to user satisfaction. Speech Commun. **74**, 12–36 (2015)
8. Schmitt, A., Ultes, S., Minker, W.: A parameterized and annotated corpus of the CMU lets go bus information system. In: International Conference on Language Resources and Evaluation (LREC), pp. 3369–3373 (2012)
9. Ultes, S., Sanchez, M.J.P., Schmitt, A., Minker, W.: Analysis of an extended interaction quality corpus. In: Natural Language Dialog Systems and Intelligent Assistants, pp. 41–52 (2015)
10. Wilcoxon, F.: Individual comparisons by ranking methods. Biom. Bull. **1**, 80–83 (1945)
11. Spirina, A.V., Sidorov, M.Y., Sergienko, R.B., Semenkin, E.S., Minker, W.: Human-human task-oriented conversations corpus for interaction quality modelling. Vestn. SibSAU **17**(1), 84–90 (2016)
12. Eyben, F., Weninger, F., Gross, F., Schuller, B.: Recent developments in opensmile, the munich open-source multimedia feature extractor. In: Proceedings of ACM Multimedia (MM), pp. 835–838 (2013)
13. Schuller, B., Steidl, S., Batliner, A.: The interspeech 2009 emotion challenge. In: Proceedings of INTERSPEECH 2009, pp. 312–315 (2009)
14. Spirina, A., Minker, W., Sidorov, M.: Could emotions be beneficial for interaction quality modelling in human-human conversations? In: Proceedings of the 20th International Conference of Text, Speech and Dialogue (TSD2017) (2017)
15. Sidorov, M., Brester, C., Schmitt, A.: Contemporary stochastic feature selection algorithms for speech-based emotion recognition. In: Proceedings of INTERSPEECH 2015, pp. 2699–2703 (2015)
16. John, G.H., Langley, P.: Estimating continuous distribution in bayesian classifiers. In: Eleventh Conference on Uncertainty in Artificial Intelligence, pp. 338–345 (1995)
17. Witten, I.H., Frank, E., Hall, M.A.: Data Mining: Practical Machine Learning Tools and Techniques. Morgan Kaufmann, USA (2011)
18. Abdi, H., Williams, L.J.: Principal component analysis. WIREs Comput. Stat. **2**, 433–459 (2010)
19. le Cessie, S., Houwelingen, J.C.: Ridge estimators in logistic regression. Appl. Stat. **41**(1), 191–201 (1992)
20. Cristianini, N., Shawe-Taylor, J.: An Introduction to Support Vector Machines and Other Kernel-based Learning Methods. Cambridge University Press, Cambridge (2000)
21. Vapnik, V.N.: The Nature of Statistical Learning Theory. Springer-Verlag, New York (1995)
22. Platt, J.: Fast training of support vector machines using sequential minimal optimization. In: Advances in Kernel Methods Support Vector Learning, vol. 3 (1999)
23. Rosenblatt, F.: Principles of Neurodynamics Perceptrons and the Theory of Brain Mechanisms. Spartan Books, Washingtion, D.C. (1961)
24. Gholap, J.: Performance tuning of J48 algorithm for prediction of soil fertility. Asian J. Comput. Sci. Inf. Technol. **2**(8), 251–252 (2012)
25. Quinkan, J.R.: C4.5: Programs for Machime Learning. Morgan Kaufmann Publishers, Burlington (1993)

26. Spirina, A., Vaskovskaia, O., Sidorov, M., Schmitt, A.: Interaction quality as a human-human task-oriented conversation performance. In: Proceedings of the 18th International Conference on Speech and Computer (SPECOM 2016), pp. 403–410 (2016)
27. Hall, M., Frank, E., Holmes, G., Pfahringer, B., Reutmann, P., Witten, I.H.: The weka data mining software: an update. SIGKDD Explor. **11**(1), 10–18 (2009)
28. Goutte, C., Gaussier, E.: A probabilistic interpretation of precision, recall and F-score, with implication for evaluation. In: Advances in Information Retrieval, pp. 345–359 (2005)
29. Rosenberg, A.: Classifying skewed data: importance to optimize average recall. In: Proceedings of INTERSPEECH 2012, pp. 2242–2245 (2012)
30. Bailey, R.A.: Design of Comparative Experiments. Cambridge University Press, Cambridge (2008)
31. Kennedy, J.J., Bush, A.J.: An Introduction to the Design and Analysis of Experiments in Behavioural Research. University Press of America, Lanham (1985)

Door Detection Algorithm Development Based on Robotic Vision and Experimental Evaluation on Prominent Embedded Systems

Alexandros Spournias, Theodore Skandamis,
Christos P. Antonopoulos$^{(\boxtimes)}$, and Nikolaos S. Voros

Technological Educational Institute of Western Greece, Patras, Greece
{alexspoul, theoskan}@cied.teiwest.gr,
{cantonopoulos, voros}@teiwest.gr

Abstract. Accurate and reliable door detection comprise a critical cornerstone for nowadays robots aiming to offer advanced features and services. At the same time, the range of robotic platforms for indoor scenarios is continuously expanding, emphasizing on the use of low cost, versatile components able to boost widespread use of such solutions. However, respective door detection algorithms have to address specific challenges such as, not relying on specialized expensive sensors, being able to offer robust and accurate operation based on commodity cameras as well as efficient execution on low resource embedded systems. Driven by aforementioned observations in this paper the design and development of a practical and reliable door detection methodology is presented based solely on typical off the shelf web-camera, while posing minimum requirements on the height or angle it is mounted. Furthermore, a critical contribution of this paper is the experimental evaluation of the developed algorithm on popular embedded systems that are based on Micro Controller Units (MCUs) commonly found on contemporary robotic platforms.

Keywords: Robotic vision · Door detection algorithm · ADLs · Embedded systems · Experimental evaluation

1 Introduction

During the last few years Robots and Robotic Platforms of various forms and types have been increasingly utilized in indoor and everyday life applications. This observation is further emphasized by the high interest respective research and development efforts have attracted by both academia and industry [1, 2]. A key reason for this continuous expansion of robots' use in common home environment relates to the drastic advancements in various research and engineering domains, effectively enabling, even low-end, robots to perform complex tasks. Furthermore, advancement in embedded systems also come to play promoting the development of low cost yet, highly functional, versatile and expandable programmable platforms.

At the end of the day, any complex operation a Robot aims to offer is based on specific fundamental, for a human being, operations which must be performed efficiently, accurately and with minimum false positives/negatives. Such a characteristic

© Springer International Publishing AG 2017
A. Ronzhin et al. (Eds.): ICR 2017, LNAI 10459, pp. 250–259, 2017.
DOI: 10.1007/978-3-319-66471-2_27

example is the capability of door detection. The importance of the door as a specific object is highlighted by the fact that it is involved in a wide range of everyday life activities of a person in an indoor environment even if, in some cases we don't notice of it because respective detection is taken for granted [3, 4]. Specifically, as a stationary object it comprises a critical landmark in any indoor environment. The position of various other objects or even the distance from a wall can be expressed in relation to a door since a door a steady, non-movable object with constant characteristics. Even more, as a mobile object (a door can open, close or reside in a range of degrees of openness), the door can be related to a wide range of activities, such as a person entering or exiting a room. At the same time an open door has different characteristics that should be taken into consideration. For example, when a robot is moving, an open door can be perceived as an obstacle or when a door is open it can pass through whereas if it is closed it cannot etc. Therefore, door detection is a crucial elementary operation for a range of complex robotic operations pertaining to Activities of Daily Life (ADLs) detection, safety and navigation [5, 6].

However, developing accurate and practical vision based door detection solutions proves to be a highly challenging task, especially when assuming zero or minimal a priori knowledge of the specific door and assumptions regarding the height and/or angle of the camera. A major reason for this challenge relates to the wide range of different and diverse types of doors encountered in various environments. Specifically, doors can have different poses, lighting situations, reflections and different specific features. Furthermore, great challenges steam from the fact that, in practical scenarios, it is common that a camera is unable to capture the whole door in a single image as well as the fact that even small differentiations regarding the distance from the door or/and the angle of the camera or the door (e.g. when a door is open) can lead to substantially different results [7].

Focusing on real deployment and operation, another aspect that is often omitted in relative efforts concerns the experimental evaluation of the execution of the detection algorithm on prominent processing units found on robots nowadays. Since one of the cornerstone objectives of such platform is power conservation, respective processing components offer relatively limited resources also aiming to contribute to minimization of energy expenditure. Therefore, it is critical to compare and objectively evaluate the efficiency and performance of any such proposed solutions to low power processing elements mainly encounter at the embedded systems domain [8].

Driven by the challenges and needs identified above the contribution of this paper is twofold. On one hand the design and development of a door detection algorithm focusing on efficiency, feasibility, low cost/complexity and resource expenditure is presented. On the other hand, a hands-on experimental evaluation is undertaken integrating and executing the proposed algorithm on prominent diverse processing environments typically encountered in contemporary robotic platforms.

The rest of the paper is organized as follows: Sect. 2 offers important background information and relative literature review. Section 3 main objective is the presentation of the design and development effort regarding the proposed algorithm, while Sect. 4 presents and analyses the efficiency of the algorithm. Section 5 focuses on the experimental performance analysis with respect to real embedded systems used in Robots and finally Sect. 6 concludes discussing the most important aspects of this work.

2   Background Information

Door detection methods have attracted considerable research interest highlighted by the wide range of different approaches proposed. The most prominent methods focus on feature extraction of the doors through an image. Such a vision-based system approach for detection and traversal of doors, is presented in Eberst et al. [9]. Door structures are extracted from images using a parallel line based filtering method, and an active tracking of detected door line segments is used to drive the robot through the door. Thus a limitation identified is that the algorithm's robustness is achieved by requiring consistency of pose estimates over time and motion of the robot.

Another method is proposed by Kragic et al. [10], where the main characteristic is that the doors are not really identified since they are imprinted on a known map. The main focus is on identifying rectangular handles for manipulation purposes. The handles are identified using cue integration by consensus. Consequently, the overall approach omits several practical and challenging realistic parameters.

Another approach widely considered, addresses the detection of doors challenge, through computer vision. Such a method was presented in 2005 from Seo et al. [11] which focused on extraction of features such as edges based on a different proposal for the vision-based door displacement behavior. In this case, the Principal Component Analysis pattern finding method is applied on images taken by the camera to identify the door.

The same year, the Muñoz-Salinas et al. [12] proposed a visual door detection system that is based on the Canny edge detector and Hough transform to extract line segments from images. Then, features of those segments are used by a genetically tuned fuzzy system that analyzes the existence of a door.

A door identification and crossing approach is also presented the 2002 in Monasterio et al. [13], where a neural network based classification method was used for both, the recognition and crossing steps. More recently, the year 2007 in Lazkano et al. [14], a Bayesian Network based classifier was used to perform the door crossing task. Doors are assumed to be open, the opening is identified, and doors crossed using sonar sensor information.

The aforementioned methods have exhibited satisfactory results, on the supposition that there aren't other objects like a door such as cabinets or libraries with the same rectangular shape. Following a respective analysis, three main challenges can be identified. The first problem is the shape because the rectangular shape of door is quite similar to other objects such as cabinets or bookcases. A second problem is the limited number of features which can be extract from a door. Finally, the third problem are the different ornamentals which can be found on different doors. All these challenges are adequately addressed by the proposed methodology focusing optimum tradeoff between accuracy and resource conservation.

Additionally, apart from the challenges posed by the characteristics of a specific door, another issue concerns the identification of the same door considering differences of visual angle views received frames (video). Such differences are created when the robot is in motion, or the door is in motion (due to opening or closing) or there are different lighting conditions etc. It is easily understood that in such cases it is considerably more difficult to identify a door compared to a static image.

3 Proposed Door Detection Algorithm Design

The proposed methodology of object detection is based on Haar-like features method [15]. The specific technic although well-known, it is typically used to identify objects characterized by high number of key features thus enabling the effective training of the algorithm. Such identification objects include faces, cars, bicycles, etc. Furthermore, aiming towards enhanced applicability the proposed approach focused on limited number of positive images (approximately 400) in contrast to at least 1500 typically required by relative approaches. In the same context, doors considered, have different characteristics on the decorative architecture. Consequently, algorithm training is stressed, because every door have different inherent characteristics (Fig. 1). In the context of the proposed technique various image processing algorithms (Fast, LBP, GBA) has been tested belonging to the Adaboost algorithm family.

Fig. 1. The differences in characteristics on the decorative architecture in each door

After several experiments, the operations of the proposed training algorithm set were segmented by using two techniques. A critical step of the followed approach concerns the handling of the photos' format, where negative photos are transformed into grayscale, whereas positives are transformed from "jpg" to "bmp" and retained in RGB forma. Following this transformation, the final classifier is extracted which proves to be highly efficient and accurate while yielding a low number of false positives recognitions.

The overall proposed procedure is depicted in (Figs. 2 and 3) and here a brief elaboration of each step is indicated.

The first stage comprised by the following steps. The first step concerns the conversion of all positive photos in "bmp" files (uncompressed) and RGB color format. In the second step all positive photos are converted to the same size while in the third step all of the negative photos are converted into in Greyscale and "jpeg" file format to save space.

In fourth step, Haar training is applied based on University of Auckland in New Zealand tools [16]. Specifically, manual annotation is conducted of positives photos while the "bg" creator tools is used for negatives photos. On every photo, annotation is made in two places (Fig. 4), one concerning the whole door contour and one focusing on the knob, and thus creating a respective file with the coordinates of annotated doors.

In fifth, and final, step for this first stage, a vector of positive images is extracted by the aforementioned file. Following the completion of the first stage of our method, as a 2nd stage we transfer the images (positives and negatives), including also the created

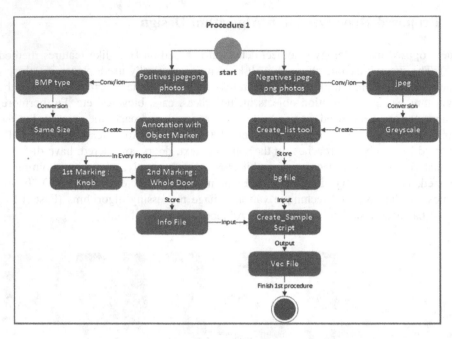

Fig. 2. 1st part of our procedure

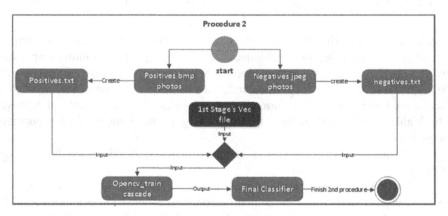

Fig. 3. 2nd part of our procedure

vector file in another Haar training set based on [17]. Finally, the positive and negative text files are created, and a training commences comprised by just 15 stages, compared to the classical automated training method requiring 22 to 25 stages. It is noted that the automated training approach even with the increased number of stages did not yielded the anticipated efficiency.

As a critical advantage of the proposed methodology, it is emphasized that while typically this process requires a three day to one-week time period, based on the

Fig. 4. Positive photos annotation

proposed approach the training lasted only approximately 45 min. It is noted that these measurements are based on experiments made on an 8 GByte RAM, 2 Core, i7 Laptop computer.

4   Evaluation of Door Detection Accuracy

For algorithm's testing and from hardware side, we used a laptop computer with i7 1st generation's processor, 8 GByte RAM DDR3 and a Logitech HD 1080P webcam. The operating system considered was the well-known Ubuntu 14.04 LTS Linux distribution, with OpenCV and Python 3.2. The implementation of the methodology was based on Yann KOËTH [18] program, from GitHub (Fig. 5).

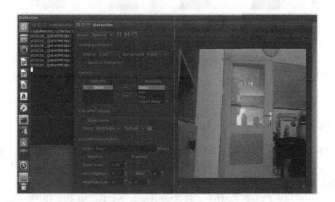

Fig. 5. Yann KOËTH detection user interface

The experiments performed considered a wide range of different interior doors. Both natural day light and artificial lighting scenarios were included as well as cases during afternoon with limited lighting and finally cases where obstacles were placed in front of the doors.

Also, a group of tests focused on different camera's tilting, ranging from 45° to 90°. Specifically, the tests which we made are described below:

(a) The first test considered ten (10) different doors with the camera mounted in front and perpendicular to them and with natural daylight. In Fig. 8, we see the results from some doors with the camera in frontal side (90˚), where there was 100% successfully recognition.

(b) The second test considered fifteen (15) different doors with the camera mounted in 45° to them and with natural daylight. In Fig. 9, we see the results where there was 80% successfully recognition. At this point, it is noted that with orientation greater than 5°, the recognition was unreliable.

(c) As a third test we performed the recognition with obstacles, in a sample of six (6) doors in which the obstruction overlapping the door incrementally reached up to 50% of the door area. The recognition proved quite reliable in all cases since positive identifications were reported in all cases.

Below in Fig. 6 we present the results from our tests.

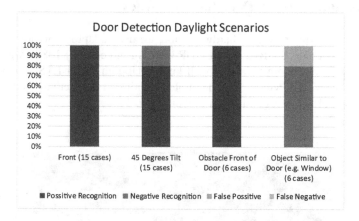

Fig. 6. Door detection test results in daylight scenarios

Explanation of types of recognition in above figure:

(a) With terminology *Positive,* we mean the recognition of the door by the algorithm.

(b) With terminology *Negative,* we mean that is not recognizing the door by the algorithm where there is a door.

(c) With terminology *False Positive* recognition, we mean the false recognition of a door by the algorithm where there is no door.

(d) With terminology *False Negative* recognition, we mean when a positive sample is misspelled as a negative (In our tests there was no such recognition).

In Fig. 7 indicative examples are depicted for the daylight experiments conducted.

The proposed approach provided quite reliable with respect to different lighting conditions since, considering the same test set, the same results were extracted considering either artificial lighting or even during afternoon with low natural light (Fig. 8).

Fig. 7. Recognition's tests indicative examples

Fig. 8. Recognition considering lit room and afternoon with natural light

Concluding it is emphasized that a negligible percentage was measured regarding false negative (i.e. a door identification was missed) while a 20% false positives was recording where positive recognition events were triggered for a shape which was not a door.

5 Experimental Evaluation of Algorithm Efficiency on Embedded Systems

Aiming to propose an efficient and practical approach with respect to nowadays Robotic systems, the detection algorithm is also tested in two embedded systems. The first system was a Raspberry pi 3 m while the second system was the PicoZed development platform from Xilinx based on the ARM Zynq processor. Both of them, although offering considerable less resources compared to an i7 based computer, they comprise prominent solutions for Robotic systems due to their low power, size, and cost characteristics combined with high degree of versatility and configurability. Of course, the same camera hardware is used in all recognition experiments.

The first and most important observation concerns the fact that there was no degradation of the recorded identification accuracy, due to the low computational resources. In fact, the results extracted concerning the identification performance are the same as the ones extracted using a resource rich computational environment.

The second observation concerns the algorithm's execution delay, where a relatively limited reduction of frame rate is observed. This is anticipated, due to the limited processing capability of the embedded hardware, yielding an output rate of the processed frames below 22 fps for Raspberry and 19 fps for PicoZed compared to ~30 frames per second recorded using the i7 laptop indicated earlier (Fig. 9).

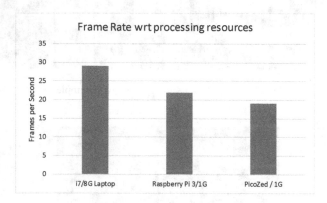

Fig. 9. Algorithm execution performance wrt prominent embedded platforms

6   Conclusions

A wide range of basic yet critical operations of modern robotic systems are based on accurate and efficient object recognition algorithms. Focusing on indoor Robot navigation and movement Door Detection comprises a critical objective due to role of a door as a moving as well as stationary area landmark. However, it is imperative that any relative methodology and algorithm takes into consideration the realistic processing capabilities of Robotic systems in order to offer a viable and realistic solution. In this paper authors address the challenge of Door Detection from multiple perspectives. On one hand a complete methodology is presented extending well established approaches offering critical advantages concerning resource conservation. In parallel proposed methodology is evaluated and validated as highly accurate and reliable in a wide range of test-objects as well as operational conditions. On the other hand, respective solution has been ported and evaluated in the context of prominent embedded systems encountered in Robotic platforms. Respective evaluation validated that the presented methodology remains highly reliable even in low resource hardware infrastructures and only a slight degradation is observed in frame rate which has zero effect on identification accuracy and robustness.

As future work, authors plan to extend the capabilities of the developed methodology so as to be highly efficient while the robot is moving and being able to identify the percentage a door is open.

Acknowledgment. This study is part of the collaborative project RADIO which is funded by the European Commission under Horizon 2020 Research and Innovation Programme with Grant Agreement Number 643892.

Open Data Access. All images used as training and test sets throughout this paper are owned by the Embedded System Design and Application Lab (http://esda-lab.cied. teiwest.gr). The training/test sets are offered to the research community for open access (https://github.com/ESDA-LAB/Door-detection) under the requirement to reference properly the current paper whenever they are published, presented or announced.

References

1. Karakaya, S., et al.: Development of a human tracking indoor mobile robot platform. In: 16th International Conference on Mechatronics - Mechatronika 2014, Brno (2014)
2. Deng, Z-H., et al.: The design and implement of household robot based on android platform. In: 2014 IEEE WARTIA, 29–30 September (2014)
3. Mahmood, F., et al.: A self-organizing neural scheme for door detection in different environments. Int. J. Comput. Appl. **60**(9) (2012)
4. Murillo, A.C., et al.: Door detection in images integrating appearance and shape cues. In: EEE/RSJ International Conference on Intelligent Robots and Systems, San Diego, USA (2007)
5. Mehmood, F., et al.: Techniques and approaches in Robocup@Home – a review. In: International Conference on Information and Communication Technologies (ICICT), 12–13 December (2015)
6. Christopoulos, K., et al.: Designing the next generation of home automation combining IoT and robotic technologies. In: 20th Pan-Hellenic Conference on Informatics (2016)
7. Hensler, J., Blaich, M., Bittel, O.: Real-time door detection based on adaboost learning algorithm. In: Gottscheber, A., Obdržálek, D., Schmidt, C. (eds.) EUROBOT 2009. CCIS, vol. 82, pp. 61–73. Springer, Heidelberg (2010). doi:10.1007/978-3-642-16370-8_6
8. Al-Shebani, Q., et al: Embedded door access control systems based on face recognition: a survey. In: 7th ICSPCS, Carrara, VIC, pp. 1–7 (2013)
9. Eberst, C., et al.: Vision-based door-traversal for autonomous mobile robots. In: Intelligent Robots and Systems, IROS 2000 (2000)
10. Kragic, D., et al.: Visually guided manipulation tasks. Robot. Auton. Syst. **40**(2–3), 193–203 (2002)
11. Seo, M.-W., Kim, Y.-J., Lim, M.-T.: Door traversing for a vision-based mobile robot using PCA. In: Khosla, R., Howlett, R.J., Jain, L.C. (eds.) KES 2005. LNCS, vol. 3684, pp. 525–531. Springer, Heidelberg (2005). doi:10.1007/11554028_73
12. Muñoz-Salinas, R., Aguirre, E., García-Silvente, M.: Detection of doors using a genetic visual fuzzy system for mobile robots. Auton. Robot. **21**(2), 123–141 (2006)
13. Monasterio, I., et al.: Learning to traverse doors using visual information. Math. Comput. Simul. **60**(3), 347–356 (2002)
14. Lazkano, E., et al.: On the use of bayesian networks to develop behavior for mobile robots. Robot. Auton. Syst. **55**(3), 253–265 (2007)
15. Viola, P., et al.: Rapid object detection using a boosted cascade of simple features. In: Conference on Computer Vision and Pattern Recognition, IEEE (2001)
16. University of Auckland in New Zealand, Tools for Haar training. https://www.cs.auckland. ac.nz/~m.rezaei/Tutorials/Haar-Training.zip
17. Thorsten Ball Opencv Haar Classifier Training. https://github.com/mrnugget/opencv-haar-classifier-training
18. Object detection Using OpenCV Haar Feature-based Cascade Classifiers, Yan KOËTH. https://github.com/xsyann/detection

Group Control of Heterogeneous Robots and Unmanned Aerial Vehicles in Agriculture Tasks

Quyen Vu[1], Vinh Nguyen[1], Oksana Solenaya[1],
and Andrey Ronzhin[1,2(✉)]

[1] St. Petersburg State University of Aerospace Instrumentation,
St. Petersburg, Russia
ronzhin@iias.spb.su
[2] SPIIRAS, 39, 14th Line, St. Petersburg 199178, Russia

Abstract. The tasks of monitoring agricultural lands using multicopters, which have higher video capture speed, higher resolution, invariance to clouds and other advantages, are considered. The aim of the research is to develop formal model and algorithms for group control of heterogeneous robotic complexes, including unmanned aerial vehicles in solving agrarian problems. Based on the analysis of existing robotic solutions in the agricultural sector, the classification of the operations is given. A formal statement of the task of controlling a group of heterogeneous agricultural robots in a certain agricultural space is formulated. We have considered the parameters of a set of cultivated lands; the number of processing agricultural objects; a set of objects of basing and storage of robotic means; a set of cultivated crops; sets of heterogeneous robots; possible options for the approach of robots from the basing area to the cultivated territory, as well as a set of resource constraints.

Keywords: Agricultural robots · Precision farming · Unmanned aerial vehicles · Multicopters · Heterogeneous robots

1 Introduction

With the development of science and technology, unmanned aerial vehicles (UAV) are increasingly being used in various sectors of the national economy. In the conduct of environmental research, UAVs provide: environmental monitoring and research; weather forecast and meteorological data collection; protection of wild animals from poaching; control over the animal population; search and rescue of people and animals; creation of maps, especially 3D-maps [1]. In agriculture, UAVs solve the problem of applying fertilizers, pesticides, etc. [2].

Agriculture accounts for the consumption of most (70%) of the world's water resources and 45% of the world's food reserves, which is produced on irrigated land, covering only 18% of the acreage [3]. Given that other national economic sectors are gradually increasing water consumption, as well as considering the current climate change projections that indicate an increase in the frequency and intensity of drought

© Springer International Publishing AG 2017
A. Ronzhin et al. (Eds.): ICR 2017, LNAI 10459, pp. 260–267, 2017.
DOI: 10.1007/978-3-319-66471-2_28

periods in the Mediterranean and semi-arid regions, the problem of monitoring water resources and their economical use in irrigation is crucial [4].

In addition, the projected global food demand by 2050 indicates that crop production should be doubled [5]. The field of plant growing was substantially developed after the Second World War as a result of the "green revolution" of the 60 s. According to the main criterion of the "blue revolution", which focuses on preserving the environment, in agriculture water management is being currently optimized to obtain the desired volume of yield per unit of water.

Within the framework of the European research program "Horizon 2020", exact agriculture is one of the priority interdisciplinary scientific directions. The American National Research Council has identified this type of agriculture as the most promising in terms of the use of information technology and robotic complexes for monitoring, obtaining data and managing crop growth, taking into account the landscape heterogeneity and variability of borders [6].

Traditional approaches of remote sensing with the placement of remote sensors on towers over fields of crops (thermal imaging, multi and hyperspectral cameras, fluorimeters, etc.) have a limited range due to the fixed position near which data are collected. Another traditional method of remote sensing is based on the use of aircraft or satellites, but temporal and spatial resolution significantly limits their effectiveness for agricultural assessments, given the very dynamic changes in vegetation with respect to the environment [7]. Also, the quality of images obtained from satellites or manned vehicles is often affected by weather conditions, so additional temporary and financial resources are expended for re-visiting at the time appropriate for shooting.

In recent years, remote sensing, based on aerial photography using UAVs, has been actively used due to technical progress, reducing costs and dimensions of sensors, the development of a global positioning system, intelligent programming systems and flight control. Improved parameters of spatial and temporal resolution of aerial photography using UAVs make it possible to extract more data on the state of the leaf cover of the crops.

The urgency of the introduction of robotic complexes in the agricultural sector is also due to social causes. The agricultural sector is characterized by heavy physical labor, monotony, heavy dependence on climatic conditions, dynamic seasonal work and other factors that negatively affect the employment of labor in the agrarian sector of the economy. In some countries of Southeast Asia, rice is one of the most important crops and the staple food. However, labor in agriculture is constantly shrinking due to the fact that the younger generation is more interested in working in offices, factories and industrial zones than in agriculture. Therefore, the activity of the agricultural sector is also declining. The most promising approach to solving this socio-economic problem is the automation of production and the use of mobile robotic complexes for agriculture.

Compared to radio-controlled aircraft, multicopters have a low cost and a low flight altitude, which ensures high resolution of images when using standard on-board cameras. In addition, due to the low flight altitude, more accurate delivery and distribution of fertilizers on agricultural land are achieved.

Considering the urgency of using UAVs for the development of the agricultural sector, in Sect. 2 we analyze the existing technical solutions with the classification of the problems being solved, and in Sect. 3 we give a formal statement of the task of managing a group of heterogeneous agricultural robots, which is the main goal of this study.

2 Review of Existing Robotic Solutions in Agriculture

Since 2000, UAVs have been increasingly used in the civil sphere and in particular for accurate farming. Initially, two types of UAVs were used for agricultural purposes: helicopters and fixed-wing aircraft. Both airplanes have a number of advantages and limitations. In the NASA projects, the Pathfinder-Plus UAV (with a wingspan of 36,3 m and weight of 318 kg, a flight time of several hours, equipped with visible and multispectral cameras to obtain images of 0,5 and 1 m per pixel respectively) was used to detect flooding and control Fertilization during the ripening of cereals in agricultural fields of coffee plantations in Hawaii in flight at an altitude of 6400 m above sea level [8]. Another development of NASA is the RCATS / APV-3 fixed wing UAV, which was used to study vineyards in California [9].

Unmanned helicopters have more sophisticated flight control systems, but provide a lower flight altitude. They are able to move in any direction and hang, maintaining a stable position in flight. These helicopters do not require special areas for take-off and landing, which is critical for standard agricultural fields. One of the latest Pheno-copter helicopter-type UAVs with a payload of 1,5 kg can fly for 30 min and perform remote measurements of the area under investigation [10].

Fixed wing aircraft use less sophisticated flight control systems and, due to a longer flight time, are able to cover a large area, but their shortcomings are a large flight altitude (as a consequence, a lower resolution of images), impossibility of hanging, and a need for a special runway [11].

The multicopter flying platform has flight characteristics similar to the helicopter, but it has more stability, maneuverability and does not require a runway. Typically, multicopters are designed from lightweight materials (carbon fiber, aluminum, fiberglass, Kevlar, etc.) with 4, 6 or 8 engines, depending on the requirements for the payload mass. Due to low cost and convenience of application, multicopters are widely used for professional and non-professional purposes. Farmers use multicopters to obtain real-time data, diagnose the state of the crop and analyze the sites that require irrigation.

Figure 1 presents a classification of the main monitoring tasks, solved with the help of UAVs on various agricultural lands. Next, consider a number of works describing the specific implementation of UAVs in solving the problems of agrarian robotics.

In [2] it is proposed to use UAVs to determine the fertility of rice using image analysis. The developed multicopter platform allows determining the fertility of rice and the required amount of fertilizers by processing images obtained from a camera installed in a multicopter.

The color chart of the leaves is used to determine the fertility of rice in the analyzed field by comparing the color of the leaves of rice plants with a list of five available green levels. The described prototype of the Quad-X multicopter platform has the following main parameters: weight 750 g, flight time 12 min, hang time 15 min,

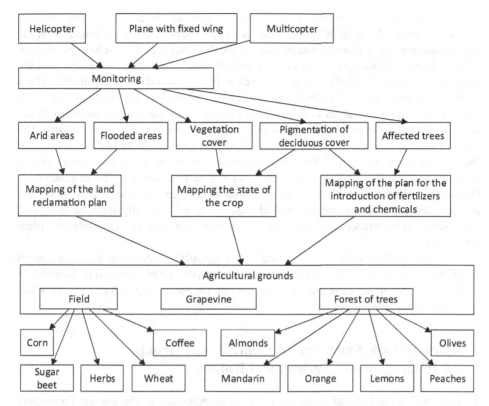

Fig. 1. Classification of tasks of agrarian robotics solved by UAVs

maximum speed 40 km/h, maximum radio distance 1000 m, price 499 euro. According to the results of the experimental check, it is revealed that the choice of the following parameters influences the efficiency of the application of the multicopter: monitoring time, flight altitude and resolution of the on-board video camera. In particular, the optimum height of the video camera was chosen – 17,37 m above the ground, and the capture time of aerial photographs – from 07.00 to 08.00 in the morning with the best light intensity for image processing. The use of multicopters for aerial photography of fields allowed farmers to reduce the amount of fertilizers by 27,1%.

In the paper [12], the use of UAVs is proposed to assess drought-prone soil and water resources needed for sustainable agricultural development. This study uses a multicopter with various non-contact sensors for remote sensing of plant conditions. The applied remote sensing technology is mainly based on the estimation of the color of leaves and other parts of plants along the wavelength and the reflection coefficient in the visible spectral range (RGB, red, green and blue) and invisible in infrared (IR) thermal radiation. As secondary indicators of plant status, the following was used: normalized vegetation difference index (NDVI); chlorophyll uptake index (TCARI); photochemical reflection index (PRI); optimized vegetative soil index (OSAVI). The NDVI index is related to the volume of plant mass and can correlate with the quality of crops, while biomass proportionately increases in parallel with photosynthesis. The photochemical reflection index (PRI) provides valuable information on the physiological state of plants when measuring the fluorescence of leaf chlorophyll from UAVs. A great advantage of UAVs in comparison with the satellite measurement of the fluorescence of leaf chlorophyll is a decrease in the altitude of their flight, which significantly increases the resolution of the image. Precise farming requires increasing the scale of captured images to extract information about the state of the plant to the level of leaves by improving the spatial and temporal resolution of remote sensing. The basis of remote sensing technology for assessing the availability of water resources of the plant is to estimate the difference between the temperatures of leaf cover, air and stomatal conductance of leaves using thermal normalized indices. The stomatal conductance of the leaves and the water potential of the leaf are useful indicators of a possible drought, and thermal normalized indices are used to analyze the variability of environmental parameters that affect the relationship between soil moisture and plant temperature.

A review of the existing UAVs, used in the agricultural sector, made it possible to identify a list of the most important operations to be performed, and also to formulate a problem statement for managing a group of heterogeneous agricultural robots while servicing some agricultural space.

3   Formal Task Statement of Control of a Group of Heterogeneous Agricultural Robots

To systematize the tasks of agrarian robotics with reference to the domain parameters that are characterized by high dynamics of processes and the limited resources available, let us next consider the task of managing a group of heterogeneous robots when servicing the workspace.

Let there be a working agricultural space S, characterized by a cortege of parameters $\langle H, P, G, C \rangle$, where $H = \{H_1, \ldots H_{i^H}\}$, $i^H \in \{1, \ldots, N\}$ is the set of cultivated lands, $P = \{P_1, \ldots P_{i^P}\}$, $i^P \in \{1, \ldots, N\}$ is the set of agricultural objects, $G = \{G_1, \ldots G_{i^G}\}$, $i^G \in \{1, \ldots, N\}$ is the set of objects of basing and storage of robotic means, $C = \{C_1, \ldots C_{i^C}\}$, $i^C \in \{1, \ldots, N\}$ is the set of cultivated crops. There is also a set $W = \{W_1, \ldots W_{i^W}\}$, $i^W \in \{1, \ldots, N\}$, which describes possible options for approach of robotic equipment from the basing area to the cultivated territory.

There are many heterogeneous robots $R = \{R_1, \ldots R_{i^R}\}$, $i^R \in \{2, \ldots, N\}$, used in this area. It is necessary to solve the problem of developing methods and algorithms for managing a group of heterogeneous robots R for servicing the workspace S in the presence of a set of resource constraints L.

Next, consider each of the parameters in more detail. Figure 1 shows an example of the layout of objects involved in the planning of agricultural robotics.

Each of the cultivated lands H_{i^H} is described by a tuple of parameters $\langle H_{i^H}^E, H_{i^H}^D, H_{i^H}^A, H_{i^H}^M \rangle$, where is $H_{i^H}^E$ is the set of coordinates of the vertices of the polygon that encompasses the boundaries of the given site; $H_{i^H}^D$ – the set of coordinates of the points of the planned regular crop plantation, grounded in its vegetative parameters;

H_{iH}^A – the cartogram of the agricultural contours [13], differing in the heterogeneous soil cover heterogeneity, its fertility; H_{iH}^M – the set of coordinates of the points that determine the trajectory of motion/flight of robotic means in the performance of agricultural tasks on the site.

As objects of basing and storage of robotics we consider only open areas for the access of ground equipment and vertical landing of UAVs. For remote sensing we are able to use external satellite and large UAVs that require a runway, but are not located on a multitude of objects of basing and storage G [14]. In this case, the object G_{iG} can be described by the set of coordinates of the vertices of the polygon enclosing the boundaries of the object. If you need to use several objects, you should consider their compact location and requirements for entrance/approaching.

The parameters of the cultivated crop C_{iC} can be represented in the form of a tuple of parameters $\langle C_{iC}^D, C_{iC}^O \rangle$, where C_{iC}^D – the description of the crop, including the parameters of planting the crop, harvesting, soil characteristics, etc. [15]; $C_{iC}^O = \left\{ C_{iC}^{O^1}, \ldots C_{iC}^{O^{iO}} \right\}$, $i^O \in \{1, \ldots, N\}$ – a set of operations planned in the process of cultivation. In turn, each operation $C_{iC}^{O^{iO}}$ is described by a tuple of parameters $\langle O_{iO}^C, O_{iO}^S, O_{iO}^D, O_{iO}^R, O_{iO}^P, O_{iO}^E \rangle$, where O_{iO} – the conditions for the beginning of the operation; O_{iO}^S – the planned time of the beginning of the operation; O_{iO}^D – the planned duration of the operation; O_{iO}^R – the resources necessary for the operation; O_{iO}^P – the set of processing agricultural objects necessary for the operation; O_{iO}^E – the set of criteria for assessing the degree of performance of an operation.

Each robot R_i is characterized by a tuple of parameters $\langle R_{iR}^B, R_{iR}^F, R_{iR}^T, R_{iR}^E \rangle$, where R_{iR}^B – the type of basing of the robot (water, land, air, space, etc.); R_{iR}^F – agricultural functions performed by the robot; R_{iR}^T – a plan of tasks set for the implementation of the robot; R_{iR}^E – a multi-criteria evaluation of the degree of implementation of the task plan.

The developed formalization for managing a group of heterogeneous robots R when servicing the workspace S must also take into account resource limitations L, including the time of operations; the available amount of equipment, including robotic, mineral and water resources; loading of processing objects, and other factors.

4 Conclusion

The urgency of the introduction of robotic complexes in the agrarian sector is caused by socio-economic reasons due to heavy manual labor and a reduction in the world's freshwater resources. Robotic means become especially popular in small-scale agriculture. Unmanned aerial vehicles are now actively being used to monitor land, map the yields of land and plan fertilization zones.

Multicopters, not requiring a runway, have high resolution and therefore high prospects for widespread use. In addition to the onboard camera, the multicopters can also be equipped with other sensory means, such as thermal imager, thermometer, gas sensors, sonar, wind speed sensors, pressure sensors, infrared and other sensors.

A distinctive feature of agrarian robotics is the relatively stable regularity of the topology of planting of cultivated crops, in contrast to other areas of application of robots, where the objects served do not have known coordinates and can move in space [16–18]. This factor somewhat simplifies the algorithmization of the trajectory of the movement of robotics in the process of performing agro-industrial operations. However, this simplification, which is characteristic of accurate farming, can be used in tasks of monitoring and cultivating land, but not when harvesting (e.g. fruit or berry crops).

The proposed formal description for solving the task of managing a group of heterogeneous agricultural robots in a certain agricultural space includes the following parameters: a set of cultivated land; a set of agricultural processing facilities; a set of basing and storage facilities for robotics; a set of cultivated crops; a set of heterogeneous robots; possible options for the entrance/approach of robotic means from the basing area to the cultivated territory, as well as a set of resource constraints.

Acknowledgments. This work is partially supported by the Russian Foundation for Basic Research (grant № 16–08–00696).

References

1. Verschoor, A.H., Reijnders, L.: The environmental monitoring of large international companies. How and what is monitored and why. J. Clean. Prod. **9**(1), 43–55 (2001)
2. Rizky, A.P., Liyantono, M.S.: Multicopter development as a tool to determine the fertility of rice plants in the vegetation phase using aerial photos. Procedia Environ. Sci. **24**, 258–265 (2015)
3. Doll, P., Siebert, S.: Global modeling of irrigation water requirements. Water Resour. Res. **38**(4), 8-1–8-10 (2002)
4. Stocker, T.F., Qin, D., Plattner, G.K., Tignor, M., Allen, S.K., Boschung, J., Nauels, A., Xia, Y., Bex, V., Midgley, P.M.: IPCC Climate Change 2013: The Physical Science Basis. Contribution of Working Group I to the Fifth Assessment Report of the Intergovernmental Panel on Climate Change. Cambridge University Press, Cambridge (2013)
5. Tilman, D., Balzer, C., Hill, J., Befort, B.L.: Global food demand and the sustainable intensification of agriculture. Proc. National Acad. Sci. U.S. Am. **108**(50), 20260–20264 (2011)
6. Anderson, K., Gaston, K.J.: Lightweight unmanned aerial vehicles will revolutionize spatial ecology. Front. Ecol. Environ. **11**(3), 138–146 (2013)
7. Jones, H.G., Sirault, X.R.: Scaling of thermal images at different spatial resolution: the mixed pixel problem. Agronomy **4**, 380–396 (2014)
8. Herwitz, S.R., Johnson, L.F., Dunagan, S.E., Higgins, R.G., Sullivan, D.V., Zheng, J., Lobitz, B.M., Leunge, J.G., Gallmeyer, B.A., Aoyagi, M., Slye, R.E., Brass, J.A.: Imaging from an unmanned aerial vehicle: agricultural surveillance and decision support. Comput. Electron. Agric. **44**(1), 49–61 (2004)
9. Johnson, L.F., Herwitz, S., Dunagan, S., Lobitz, B., Sullivan, D., Sly, R.: Collection of ultra-high spatial and spectral resolution image data over California vineyards with a small UAV. In: International Symposium on Remote Sensing of Environment, Honolulu, HI, 10–14 November 2003

10. Chapman, S.C., Merz, T., Chan, A., Jackway, P., Hrabar, S., Dreccer, M.F., Holland, E., Zheng, B., Ling, T.J., Jimenez-Berni, J.: Phenocopter: a low-altitude, autonomous remote-sensing robotic helicopter for high-through put field-based phenotyping. Agronomy **4**, 279–301 (2014)
11. Zarco-Tejada, P.J., González-Dugo, V., Williams, L.E., Suárez, L., Berni, J.A.J., Goldhamer, D., Fereres, E.: A PRI-based water stress index combining structural and chlorophyll effects: assessment using diurnal narrow-band air-borne imagery and the CWSI thermal index. Remote Sens. Environ. **138**, 38–50 (2013)
12. Gagoa, J., Douthe, C., Coopmanc, R.E., Gallegoa, P.P., Ribas-Carbo, M., Flexas, J., Escalona, J., Medrano, H.: UAVs challenge to assess water stress for sustainable agriculture. Agric. Water Manag. **153**, 9–19 (2015)
13. Afanas'ev, R.A., Ermolov, I.L.: Future of robots for precision agriculture. Mechatron. Autom. Manag. **12**, 828–833 (2016)
14. Jakushev, V.P., Petrushin, A.F.: Possibilities for estimation of reclaimed agricultural land quality given by accumulation and processing information from remote sensing. Agrophysics **2**(10), 52–58 (2013)
15. Sidorova, V.A., Zhukovsky, E.E., Lekomtsev, P.V., Yakushev, V.V.: Geostatistical analysis of soil characteristics and productivity in the field experiment on precise agriculture. Agrochem. Fertil. Soils **8**, 879–888 (2012)
16. Vatamanjuk, I.V., Panina, G.J., Ronzhin, A.L.: Modeling of robotic systems' trajectories in spatial reconfiguration of swarm. Robot. Tech. Cybern. **3**(8), 52–57 (2015)
17. Krjuchkov, B.I., Karpov, A.A., Usov, V.M.: Promising approaches for the use of service robots in the domain of manned space exploration. SPIIRAS Proc. **32**, 125–151 (2014)
18. Motienko, A.I., Tarasov, A.G., Dorozhko, I.V., Basov, O.O.: Proactive control of robotic systems for rescue operations. SPIIRAS Proc. **46**, 169–189 (2016)

Tracking of Warehouses Robots Based on the Omnidirectional Wheels

Alexey Zalevsky[(✉)], Oleg Osipov, and Roman Meshcheryakov

Tomsk State University of Control Systems and Radioelectronics, Tomsk, Russia
Zalevskiy.aleksey@gmail.com, ems2009@mail.ru,
nauka@tusur.ru

Abstract. The paper deals with the design robots platforms for warehouse. The experiment investigated the robot modes. Platform contains four omnidirectional wheels and standardized electronic control system. Proposed upgrade ways platform.

Keywords: Robot · Warehouse · Platforms · Manipulator · Rotation wheel · Control systems

1 Introduction

It is necessary in enterprises, whose activities forced to have a number of different parts, components and other production elements, have a warehouse. More items need to be stored; the more must done strictly storage organization. Otherwise, the process of finding the required type of product in stock will require an indefinite amount of time. Another requirement for the storage organization is density. The more compact storage, the less time will spent on search and transportation, and the more units can be stored in the same area. Often, stores do not do with dense organization, due to fact that in such cases there is not enough space for performing transportation of cargo, to perform maneuvers such as turning and turning. In those cases where the transport is not possible to get to the object storage space, this work has to do staff.

The existing unsolved problems of omnidirectional platforms are described in many scientific papers. In the work "2D path control of four omni wheels mobile platform with compass and gyroscope sensors" authors create design of the platform to be sure that four wheels can always contact with uneven ground surface for avoiding wheel slipping or idle running [1]. Other researchers have paid attention to the importance of the speed value of rotation of the wheels. They in their paper "Trajectory tracking for omnidirectional mobile robots based on restrictions of the motor's velocities" propose an algorithm that combines the restriction on the motor's velocities and the kinematic model of Omnidirectional mobile robots to improve the trajectory's following [2]. Proper work with the drivers is very important because the accuracy of the platform positioning depends on this. The new approach in a trajectory-planning controller, which are presented in the article "model-based PI-fuzzy control of four-wheeled omnidirectional mobile robots" results in more precise and proper outputs for the motion of four-wheeled omnidirectional mobile Robots [3].

© Springer International Publishing AG 2017
A. Ronzhin et al. (Eds.): ICR 2017, LNAI 10459, pp. 268–274, 2017.
DOI: 10.1007/978-3-319-66471-2_29

The most famous industrial omnidirectional platform for cargo transportation is KUKA omniMove (Fig. 1). This company makes a mobile platform for heavy cargo. The KUKA omniMove can be controlled manually or move autonomously. Despite its enormous size and payload capacity, it navigates safely, moving virtually independently. Specially developed wheels allow the mobile heavy-duty platform to move in any direction – even from a standing start. The sophisticated navigation system KUKA Navigation Solution ensures autonomous maneuvering without risk of collision and without requiring artificial floor markings. The KUKA omniMove can be freely scaled in size, width and length within a modular system – in accordance with the requirements.

Fig. 1. KUKA omniMove platform

There is one mobile learning arduino kit (Fig. 2). It includes four coreless 12 V DC motors with encoders, arduino microcontroller and I/O expansion on board; it is programmable with the open source Arduino language. Its chassis is made of Aluminum-alloy.

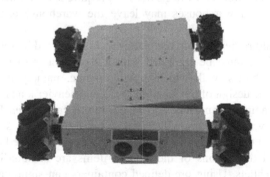

Fig. 2. Arduino kit platform with movable frame

This Mecanum wheel mobile Arduino robotics car can be made to move in any direction and turn by varying the direction and speed of each wheel. The platform rear wheels are mounted in a particular way, so that the suspension structure ensures that all four wheels can adhere to the ground, even when the ground is uneven.

2 Platform

We suggest the creation of universal platform. Driving directions can be take all possible values that is our platform can move any direction and is capable of performing turns in space. When we have such opportunities, we will not need to perform the preliminary maneuvers to find permitted direction. There are several ways to implement Omni movement. The most common way is using omnidirectional wheel in a view of its ease of technical implementation. Traction vector of each $\overrightarrow{Tv}$ of four wheels is directed to $45°$ ($\overrightarrow{D}$ – fixed vector direction of traction) to the plane of rotation. Speed of rotation ϑ and tire-to-surface coefficient α determine value of traction vector:

$$\overrightarrow{Tv} = \alpha \cdot \vartheta \cdot \overrightarrow{D}. \tag{1}$$

The vector summarize of movement $\overrightarrow{M}_\Sigma$ may give any direction platform for arbitrary speed of the wheels:

$$\overrightarrow{M}_\Sigma = \sum_{i=1}^{4} \overrightarrow{Tv}_i. \tag{2}$$

Robotic omnidirectional platform is ideal movement tool in compact warehouses. It does not require extension of corridors between the shelves stock due to lack of the need additional space for maneuvering movement. Place shelves so close to each other as far as permitted items stored on these shelves. Dimensions platform are performed like the minimum width of the corridor. It realized a significant gain in compactness of warehouse where storage items are small and for transporting them does not require wide corridors. This platform on wheels does not require special rails, sufficiently hard, level surface. Robot of warehouses may leave the warehouse at any point to load luggage.

Each item should retrieved from the platforms luggage and be moved to permanent storage. The platform must be equipped with a manipulator to perform this operation (Fig. 3). The objects, with which it will operate, namely the shape, size and weight, determine the design of the manipulator. Than heavier loads the more powerful platform and the manipulator must be performed. In this case, for each type of load, it is required to have a way of capture which will safely move it to a specific location.

If there is a need the security control then it is set camera aimed at the manipulator motion. Shape and dimensions of the storage items are standardized by means of special cases and pallets. Using pre-defined containers can significantly simplify and speed up the process of loading and unloading robotic platforms.

Fig. 3. Platform with manipulator

3 Test

If there is a need the security control then it is set camera aimed at the manipulator motion. However, often shape and dimensions of the storage items are standardized by means of special cases and pallets. Using pre-defined containers can significantly simplify and speed up the process of loading and unloading robotic platforms [8–11].

Currently electromechanical test platform structure is assembled and minimally equipped with the necessary electronics (Fig. 4). Frame sizes can be changed easily, as the structure is made maximum multipurpose.

Fig. 4. Test platform structure

Program for manual control of platform was written and tests were performed. It became clear that there are three independent type of movement. They are presented in Fig. 5 and corresponding rotation directions of wheels.

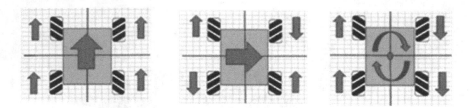

Fig. 5. Independent types of movement

If combine different type of movement at different proportion then we can get any trajectory as it is presented in the Fig. 6. Full trajectory of the robot is shown in Fig. 7.

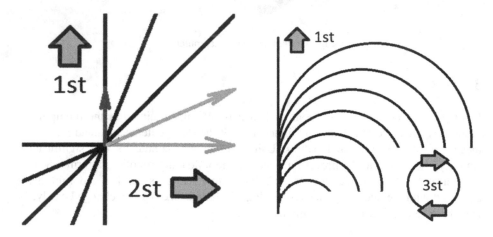

Fig. 6. Combination of movement

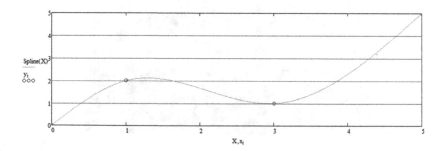

Fig. 7. Robots trajectory

In addition, some critical moments in the assembled mechanism were identified. Firstly, as the final direction of movement takes into account the contribution of each wheel, the loss of contacts any of the wheels to the surface, the initial trajectory is broken. Loss of contact occurs when all four wheels are on the same plane, and the surface is a curve at which the robot rides.

The second critical moment is the free running wheel rotation. Encoder cannot trace these free rotations because it is located by motor shaft. Therefore, the more free rotation, the more error in the initial direction of motion. In constant motion, this problem does not contribute to the error of the trajectory as the wheels rotate always in the same direction and back play and a change of the belt tension direction only happens when you change the direction of rotation of the motor shaft. Other critical moments can explored after removing the first two, as they are the most affect the right movement.

4 Conclusion

To ensure permanent contact of all 4 wheels with a rough surface, it needs to rework the suspension or make a movable frame along the axle. The independent suspension of all wheels is expensive way to solve this problem so for this robot it is not advisable to use it. Performing the movable frame along the axle means that the front and rear pair of wheels will always be linked to the surface. Between them, these parts are connected by the axis that allows platform is bent taking the shape of the surface (as in the Fig. 2).

Reduction of a free running wheel is implemented by performance by more rigid hitch of the mechanism of transfer of rotation from an engine shaft on a wheel [12–15].

Acknowledgements. The given paper is completed with the support of the Ministry of Education and Science of the Russian Federation within the limits of the project part of the state assignment of TUSUR in 2017 and 2019 (project 2.3583.2017).

References

1. Huang, S.J., Shiao, Y.W.: 2D path control of four omni wheels mobile platform with compass and gyroscope sensor. Sens. Actuators A: Phys. **234**, 302–310 (2015). doi:10.1016/j.sna.2015.09.012
2. Conceição, A.S., Moreira, A.P., Costa, P.J.: Trajectory tracking for omni-directional mobile robots based on restrictions of the motor's velocities. IFAC Proc. Vol. **39**(15), 121–125 (2006). doi:10.3182/20060906-3-IT-2910.00022
3. Hashemia, E., Jadidib, M.G., Jadidi, N.G.: Model-based PI–fuzzy control of four-wheeled omni-directional mobile robots. Robot. Auton. Syst. **59**(11), 930–942 (2011). doi:10.1016/j.robot.2011.07.002
4. Antonov, A.: Types of wheels of mobile robots. http://robotosha.ru/robotics/wheel-types-mobile-robots.html. Accessed 15 Feb 2017
5. Osipov, Y.M., Izotkina, N.Y.: Modern Problems of Innovation: The Manual. TUSUR, Tomsk, 140 p. (2012). https://edu.tusur.ru/publications/1056
6. Dovbnya, N.M., Kondratyev, A.N., Yurevich, E.I.: Robotized Technological Complexes. Mechanical Engineering, Department of Leningrad, Leningrad (1990)
7. Bespalov, R.S.: Transport logistics. Latest technology of building an effective delivery system. M.: Vershina, 384 p. (2007)

8. Kostyuchenko, E., Gurakov, M., Krivonosov, E., Tomyshev, M., Mescheryakov, R., Hodashinskiy, I.: Integration of bayesian classifier and perceptron for problem identification on dynamics signature using a genetic algorithm for the identification threshold selection. In: Cheng, L., Liu, Q., Ronzhin, A. (eds.) ISNN 2016. LNCS, vol. 9719, pp. 620–627. Springer, Cham (2016). doi:10.1007/978-3-319-40663-3_71

9. Khodashinsky, I.A., Meshcheryakov, R.V., Anfilofiev, A.E.: Identification of fuzzy classifiers based on weed optimization algorithm. In: Creativity in Intelligent Technologies and Data Science First Conference, CIT&DS 2015, Proceedings, pp. 216–223 (2015)

10. Khodashinsky, I.A., Zemtsov, N.N., Meshcheryakov, R.V.: Construction of fuzzy approximators based on the bacterial foraging method. Russ. Phys. J. **55**(3), 301–305 (2012)

11. Osipov, O.Y., Osipov, Y.M., Meshcheryakov, R.V.: Active driveline as an element of cyberphysical system. Proc. High. Educ. inst. Instrum. **59**(11), 934–938 (2016)

12. Vasiliev, A.V., Kondratyev, A.S., Gradovtsev, A.A., Dalyaev, I.Y.: Research and development of design shape of a mobile robotic system for geological exploration on the moon's surface. SPIIRAS Proc. **45**(2), 141–156 (2016). doi:10.15622/sp.45.9

13. Kryuchkov, B.I., Karpov, A.A., Usov, V.M.: Promising approaches for the use of service robots in the domain of manned space exploration. SPIIRAS Proc. **32**(1), 125–151 (2014). doi:10.15622/sp.32.9

14. Motienko, A.I., Tarasov, A.G., Dorozhko, I.V., Basov, O.O.: Proactive control of robotic systems for rescue operations. SPIIRAS Proc. **46**(3), 169–189 (2016). doi:10.15622/sp.46.12/

15. Ronzhin, A., Saveliev, A., Basov, O., Solyonyj, S.: Conceptual model of cyberphysical environment based on collaborative work of distributed means and mobile robots. In: Ronzhin, A., Rigoll, G., Meshcheryakov, R. (eds.) ICR 2016. LNCS, vol. 9812, pp. 32–39. Springer, Cham (2016). doi:10.1007/978-3-319-43955-6_5

Open-Source Modular μAUV for Cooperative Missions

Vladislav Zanin, Igor Kozhemyakin[✉], Yuriy Potekhin,
Ivan Putintsev, Vladimir Ryzhov, Nickolay Semenov,
and Mikhail Chemodanov

St. Petersburg State Maritime Technical University, SMTU, Saint Petersburg,
Russia
1861vp@mail.ru

Abstract. This paper examines the results of work undertaken by Saint-Petersburg State Marine Technical University (SMTU) as part of a project involving research into the creation of a multi-agent sensory network based on marine robotic platforms (MRP). In the context of the works mentioned, creation of a micro autonomous underwater vehicle (μAUV) is considered. High maneuverability, the availability of a modular architecture and functions of the group use are required for the developed μAUV. The article describes steps of the vehicle creation: concept development, modeling, design, construction, simulation of the implementation of various vehicle group mission scenarios (Solving the task of monitoring the seabed area). A standard chassis which allows the assembly of all systems, units and mechanisms of the device in minimum dimensions, has been established. Within the project, a software and hardware architecture of the information system of the vehicle was developed, as well as a model of interaction between the μAUV, the wave glider and control center. The work results in two full-scale experimental samples of μAUV capable of working in a group together with the wave gliders, performing the functions of a repeater signal through the environmental boundary, and simulation results of implementing the mission by a group of developed μAUV.

Keywords: Marine multi-agent sensor network · Marine robotic platform · Micro autonomous underwater vehicle · Wave glider-retranslator · Group missions

1 Introduction

In 2016 in the SMTU an initiative project was carried out in the development of research [1–4] on the creation of a multi-agent sensory network of marine robotic platforms.

Within this project, a mathematical and simulation model for the interaction of homogenous/heterogeneous robots was developed, micro autonomous underwater vehicle (μAUV) samples were made, functioning as underwater network agents, as well as control algorithms were transferred from the simulation model to the real controlled objects. As a result, a complex of full-scale experimental samples of

© Springer International Publishing AG 2017
A. Ronzhin et al. (Eds.): ICR 2017, LNAI 10459, pp. 275–285, 2017.
DOI: 10.1007/978-3-319-66471-2_30

autonomous unmanned vehicles was created, consisting of a wave glider, acting as an environmental gateway node between an underwater and a surface environment, and a group of μAUV, which is the main task performer of searching for objects and monitoring of the underwater environment.

The innovative aspect of the implemented project is the development of a modular μAUV, able to independently detect objects on the seabed, as well as unification of an open group of different types of robots into a self-organizing network that is stable both to a change in the number of interacting robots and to a low speed of interaction, depending on the mutual arrangement of robots among themselves and on hydrology.

Also the following performance characteristics were required: depth up to 50 m, battery lifetime – not less than 3 h, weight - not more than 10 kg, maximum speed – not less than 3 knots.

Vehicle payload assumed to consist of the main part, presented in all the configurations (CTD sensor and underwater vision system), and some optional parts (side scan sonar, an electromagnetic field detector etc.).

Mathematical modeling of the hydrodynamic and strength characteristics of the vehicle was performed, two full-scale experimental μAUV models were designed, fabricated and assembled, software and hardware architecture of the information system was worked out, and research of vehicles group application was carried out.

Below is a brief description of the main stages of the project.

2   Vehicle Architecture

For μAUV in the basic configuration, the following module composition was defined (Fig. 1): nosecone module, fore variable buoyancy engine (VBE) module, fore through-hull tunnel thruster for vertical motion (TT), electronics and hydro acoustics module, power supply/replacement battery module, aft through-hull tunnel thruster for vertical motion (TT), aft variable buoyancy engine (VBE) module, main thrusters module.

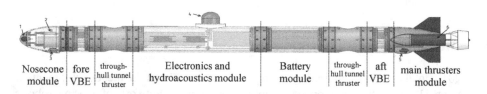

| Nosecone module | fore VBE | through-hull tunnel thruster | Electronics and hydroacoustics module | Battery module | through-hull tunnel thruster | aft VBE | main thrusters module |

Fig. 1. μAUV modular architecture (1 – forward looking ultrasonic sonar, 2 – temperature, depth, pitch & roll sensor, 3 – fore light, 4 – hydro acoustic modem antenna, 5 – aft light, 6 – main thruster module)

3 Vehicle Functional Systems

Variable buoyancy engine (VBE) provides µAUV buoyancy adjustment for maintaining its position at a given depth. VBE is a piston driven mechanism, changing dry volume of the vehicle. The vehicle has two VBEs –fore and aft VBE modules.

A **sonar system with ultra-short base location** (USBL) modem is used for positioning and navigation. Modem EvoLogics S2C 42/65 USBL was used [7].

Surface radio communication system is built on the basis of WiFi module D-Link DWA-137/A1 with a plug-in external antenna.

Range detection system (Forward looking ultrasonic sonar) was developed and manufactured on the basis of the overall dimensions allocated for a piezoelectric antenna, placed in a vehicle nosecone module, and necessary technical characteristics [7]. Maximum range is 10 m, antenna is circular hemispherical with a radius of 20 mm, located objects must be more than 50 mm, central frequency is 460 kHz. The width of the transducer directivity characteristic is about 70°.

Vehicle CPU module is presented by board Beaglebone Black [7], operating under Linux OS. The main advantages of the board are compactness and a large number of integrated peripherals.

Temperature, depth, pitch & roll sensor, located in the nasal module, is built on the basis of the OpenROV IMU/Compass/Depth Module [7]. This module is well proven in use on the ROV previously developed by the authors [7]. Overall size of the selected module is comparable with the leaders according to the declared characteristics and provides the necessary technical characteristics of the vehicle.

Underwater vision system includes camera module, fore and aft lights. As a module of the video camera, the board Aptina RB HD Camera Cape for BeagleBone Black A0-01 [7] was used. As a fore and aft lights of the underwater vision system OpenROV External Light Cube models were selected, [7].

Power supply system µAUV built on Li-ion rechargeable batteries. As a basic element, the element Samsung 18650 was used, which, due to its geometric characteristics, allowed to fill the battery module space quite efficiently. To obtain the required voltage and capacity, 28 such elements are used. The battery assembly provides a voltage of 14.8 V and a capacity of 18.2 Ah, which gives the required power consumption for 3 h.

Propulsion system of the vehicle is based on two main thrusters CrustCrawler 400HFS [7], placed in a vehicle tailcone module.

Module of through-hull tunnel thruster for vertical motion is based on the brushless motors Turnigy Aerodrive DST-700 [7] with a corresponding driver.

The use of the above components determined the overall performance of the vehicle module and of the assembled µAUV.

4 Simulation of the Main Vehicle Characteristics

For preliminary hydrodynamic analysis, assessment of hydrodynamic forces and moments in longitudinal-vertical and longitudinal-horizontal planes was carried out. The Ship Dynamic Software (SMTU) generates hydrodynamic characteristics on the basis of information about the shape of the object.

Based on the results of the calculations, it was concluded that the dynamic stability of the rectilinear μAUV motion is ensured in both the vertical and horizontal planes, which is an extremely important condition for a high-quality automated control of a moving object in a target tracking mode, searching, monitoring of the water area and vertical self-balancing.

Required level of damping components hydrodynamic characteristics is completely provided by propellers nozzles, without requiring the installation of additional tail empennage elements. As to the positional hydrodynamic characteristics, it can be noted, that they, in general, are a completely traditional for objects of this type.

An important element is the depth control system of the vehicle. On the developed vehicle two systems are used, allowing to control its depth of immersion: tunnel thrusters for vertical motion and VBE. The method used for depth control is a combination of proportional-derivative (PD) controller for propulsion and steering complex and PID controller for VBE. The aim of the method is to combine vehicle fast-moving to a given depth with low power consumption in a steady state. The flowchart of the algorithm is given on the Fig. 2:

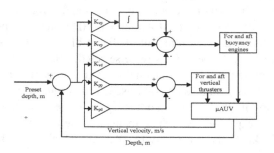

Fig. 2. Block diagram of depth control algorithm

As a basis for PID controller implementation the classical construction principle with a modification of the calculation of the integral component was adopted:

$$\begin{cases} F(t) = K_p e(t) + K_d v(t) + \sum_{\tau=0}^{t} I_q \\ I_q = K_i T \dfrac{V}{a \cdot e(t)} \end{cases}, \partial e \begin{cases} V = 0, |v(t)| \geq b \\ V = 1, |v(t)| < b \end{cases}, \tag{1}$$

where K_p, K_d, K_i – PID controller coefficients, $e(t)$ – residual of the parameter to be stabilized, $v(t)$ – rate of change of the parameter to be stabilized, T – integration period, a, b – customizable coefficients.

It follows from Fig. 3 that through-hull tunnel thrusters for vertical motion operate for a short time, providing a vertical speed at the initial stage of functioning, VBE starts to work in parallel, and within 15 s the vehicle stabilizes at a new depth without help of thrusters for vertical motion and additional energy costs.

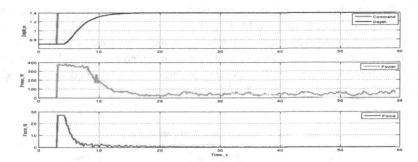

Fig. 3. Graphs of the main depth control algorithm parameters

Result of manufacturing μAUV (in its basic configuration) is shown in Fig. 4.

Fig. 4. μAUV appearance (in the basic configuration)

5 Software and Hardware Architecture of the Vehicle Information System

Functionality of the software and hardware complex is divided into two levels - upper and lower. Functionality (subroutines) of the low-level functionality is responsible for the basic elements of the device's operation ensuring trouble-free operation of the device, protection from over-reaching of acceptable ranges of operation, as well as the functioning of high-level subroutines. The main difference between the functionality of the lower and upper levels is that the lower layer itself is not responsible for the solution of the problem, instead it facilitates the performance of high-level tasks.

The basic functions (subroutines) include: holding the preset depth, transitioning between different layers of depth, remaining within specified coordinates, avoiding collisions with obstacles directly in the vehicle's path, finding the vehicle's own coordinates, movement to a point with given coordinates, recognition of objects on the sea floor.

High-level functionality (subprograms) includes: construction of a survey route by the perimeter of the water area, informing the control center about changing the status of the mission (the object was found in the selected zone, the object was not found), changing the operating mode according to the control center command, returning the vehicle to base.

The hardware architecture of the μAUV information network is based on the "star" topology, in which the Beaglebone black processor plays the role of the network central module, and the peripheral elements are the sensors and actuators. Communication between the central module and peripheral devices is provided through various communication channels: SPI, I2C, PWM, UART.

The electronics and hydroacoustics module is the most loaded with electronics. There are: a hydro-acoustic modem, central processor, video camera and drivers for all brushless motors in the device. The modem antenna is also attached to this module, Fig. 5.

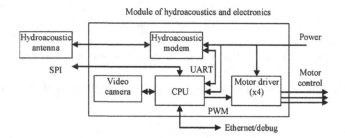

Fig. 5. Functional diagram of the electronics and hydroacoustics module

In the stern module there are two main thrusters and a stern light source. The software of the device is based on the ROS framework [7].

6 Data Flow Repeater of Data Flow from μAUV to the Control Center

Retransmission of the data flow between the μAUV and the control center is carried out over a mixed hydroacoustic/radio channel using the wave glider repeater located in the mission area of the μAUV grouping.

Retranslation of data between μAUV and control center can be constructed either by full retransmission of all data between μAUV and control center, or by preliminary processing and reduction of the transferred data.

7 Control Center

The control center post interacts primarily with the surface module of the wave glider repeater. The exchange of information between the surface module and the control center is carried out via WiFi.

To determine the coordinates of underwater vehicles, the surface wave module of the wave glider must know its own coordinates and course, for which in open water conditions it is supposed to use a GNSS receiver of the coordinates and course, and under test conditions in the experimental pool - an ultrasound system for positioning inside the room.

Software for the control center is based on NS2 by Nonius Engineering [7], and provides the following basic capabilities: display of positions: underwater vehicles, surface wave module of a repeater glider, search area, the ability to specify the search area, sending commands to the mission execution unit (return to the base, assigning a new area, etc.).

8   Group Application of Vehicles

Upon completion of the test phase, it is proposed to use group vehicles with multi-beam echo sounders as a payload. Consider the effectiveness of the application of a group of such similar μAUVs for solving the problem of monitoring the sea surface of a complex shape. The communication between μAUV, mutual positioning and control is limited by relatively low-speed modems and limitations on the range of communication. Therefore, the task of managing the μAUV group is to coordinate the actions of individual μAUVs in such a way as to achieve high efficiency in solving a common problem with a limited information transfer rate between the μAUV. The question remains - how does the number of μAUVs in the group affect the effectiveness? How many μAUVs are needed to effectively solve the problem?

Several variants of the solution of this problem were modeled:

a. Use of one AUV with a predetermined sequence program;
b. Use of the AUV group moving in a line to increase the width of the investigated area in one pass;
c. Use of the AUV group with the simplest group management algorithms, such as "flock" or "swarm";
d. Use of the AUV group with algorithms of multi-agent systems (MAS).

The simulation was carried out in the Matlab package. When using a multi-beam sounder in one pass of one AUV, it is possible to survey an area of no more than 10 m at a depth of 8−12 m. Thus, one AUV, moving along the optimal path from the point of view of the number of turns of the trajectory at a speed of 3 m/s, by 98 h. An example of a territory survey by a single robot is shown in Fig. 6a:

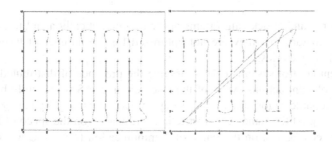

Fig. 6. An example of a trajectory by (a) a single robot (b) two robots side-by-side

Figure 6a shows that the robot moves in a "snaking" path, since it is this type of motion that ensures the least overlapping of the coverage areas when the robot examines an arbitrary territory.

The movement of several AUVs side-by-side sharply increases the width of the surveyed space by the number of AUVs in the group, but at the same time complicates the maneuvers. Thus, when two AUVs moved in the same rank, the time required for the survey was 51 h, that is, almost twice as effective as using one AUV. But with the increase in the number of AUVs to 10, the efficiency drops sharply – about 36 h is obtained during the simulation. No information exchange is required between the AUVs, only mutual positioning is sufficient.

An example of a survey of the territory by two AUVs is shown in Fig. 6b.

Figure 6b shows that in the process of maneuvering, one robot is forced to wait until the second takes the required position for further side-by-side movement. This expectation reduces the effectiveness of this method, especially when a lot of robots are involved in the line and the waiting time is significant compared to the group's overall working time.

Using the simplest control algorithms that do not require trajectory planning and for which only mutual positioning and marking of the investigated areas is sufficient. The "swarm" type algorithm can be implemented in the form of a desire to take the nearest unexplored site on the map as far from other AUVs. More complex algorithms suggest the development of a non-linear function of the "cost" of each unknown site, when the goal of each AUV is to take unexplored objects of maximum cost faster than other AUVs. Examples of the trajectories of motion along the described algorithms are given in Fig. 7.

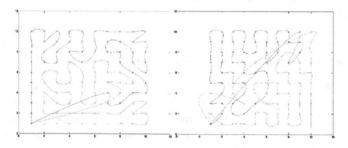

Fig. 7. An example of the trajectory of two robots using the "swarm" and "flock" algorithm

The "swarm" and "flock" algorithms allow to automatically adapt to any shape of the surveyed territory, but they are not optimized for the number of turns and intersections of the trajectories.

Multi-agent control systems (MAS) assume the presence of distributed computing and planning of joint actions. Most of the research on MAS is currently based on the search for a reliable algorithm for ensuring agreement among agents performing a single task [8, 9]. At the same time, restrictions are placed on the speed of information exchange between agents, on the range of communication (not all agents can directly inform everyone else about the information, before some agents the message must be delivered along the chain) and positioning accuracy. Such algorithms are called "consensus problems" in distributed systems [9, 10] and are implemented in open source systems DELLPHIS, Actors, Blackboards and Contract Net Protocol. The protocol of exchange between the AUVs is also standardized and is called JADE.

An example of a survey trajectory for two robots using the MAC algorithm is shown in Fig. 8. It is clear that robots plan their work, their trajectories do not intersect, the number of turns is less than in other algorithms, and none of the robots stand idle in anticipation of another.

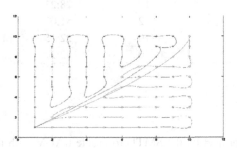

Fig. 8. An example of a motion trajectory of two robots according to the MAS algorithm

The following mission execution and operational control protocol is used:

1. Each agent must know the whole scenario of the execution of the application;
2. Each agent must know the distribution of performers according to the script's work;
3. Each agent should know what kind of actions and what agents are being performed at the moment.

If conditions for the script execution change, the scenario is changed using the Contract NET Protocol (CNP).

As group control efficiency, the criterion of loading time of each μAUV was used for the entire period of the group's work:

$$E = \frac{T_1}{N \cdot T_N} \cdot 100\%, \qquad (2)$$

where N is the number of members of the group; T_1 – the time during which one robot can perform the entire task independently; T_N – the time for which the task is performed by a group of N members. Results are shown in Table 1.

As can be seen in Table 1, using MAS algorithms makes it possible to use larger groups of robots more efficiently. Other algorithms considered with an increase in the number of robots are much more losing the efficiency of using each robot than with the use of MAS. This is due, apparently, to the possibility of trajectory planning by each robot, taking into account the movement of the other participants and efficient exchange of current information between the robots.

Table 1. Results of modeling the survey of the territory by the AUV group

Management structure	AUV quantity	Survey time, h	Effectiveness, %
1. One AUV with the given program	1	98:00	100%
2. Movement in a line	2	51:12	95%
	10	35:56	28.8%
3. Algorithm "swarm"	2	50:38	96.7%
	10	32:47	29.9%
4. Algorithm "flock"	2	53:04	92.3%
	10	28:23	34.3%
5. MAS	2	51:07	95.9%
	10	22:15	44.0%
	20	12:27	39.3%

9 Conclusion

While implementing project by the SMTU to create μAUV with the function of group control, the following tasks were solved:

- developed a conceptual model of the joint use of the μAUV group and the wave glider repeater within the multi-agent sensor network;
- μAUV appearance and modular architecture were developed;
- computational estimation of the hydrodynamic, strength, and acoustic, energy characteristics of μAUV is carried out;
- constructive elements of the μAUV hull structure, parts and mechanical parts of the apparatus systems were manufactured;
- two μAUV are assembled;
- basic software for μAUV, wave glider and control center architecture are developed with the possibility of further development and refinement;
- technical solution has been developed, which provides for carrying out experiments with the MRP complex in a towing tank;
- comparative analysis of various μAUV group application algorithms for the water area survey is performed;
- plan for further testing of μAUV in the SMTU towing tank is developed.

References

1. Research for support the creation of the information and measuring system based on the "glider" type autonomous unmanned underwater vehicles. Technical reports on 1,2,3 stages, R&D Department, SMTU, № 01201280856 (2012–2014). (in Russian)
2. Experimental studies for support the creation of the wave glider-type autonomous unmanned vehicles. Technical report, R&D Department, SMTU (2015). (in Russian)

3. Kozhemyakin, I.V., Rozhdestvensky, K.V., Ryzhov, V.A.: The wave gliders as the marine global information and measurement system elements. Materialy desyatoi nauchno-prakticheskoi konferencii «Perspektivnye sistemy i zadachi upravlenya» – Taganrog, SFedU, pp. 101–112 (2015). (in Russian)
4. Kozhemyakin, I.V., Rozhdestvensky, K:V., Ryzhov, V.A.: Development of the technical platform of the global maritime information and measuring system based on the glider-type autonomous unmanned vehicles. In: Rossiyskie innovacionnye tehnologii dlya osvoeniya uglevodorodnyh resursov kontinentalnogo shelfa, pp. 91–108 (2016). (in Russian)
5. Zanin, V.Y., Kozhemyakin, I.V., Potekhin, Y.P., Putintsev, I.A., Ryzhov, V.A., Semenov, N.N., Chemodanov, M.N.: Development of micro autonomous underwater vehicles with control group function. Izvestiya SFedU. Eng. Sci. **1–2**, 55–73 (2017). (in Russian)
6. Semenov, N.N.: Comparison of different types of probing signal for the active sonar. Informatsionno-upravlyayucshie sistemy **9**(41), 23–28 (2008). (in Russian)
7. Kozhemyakin, I., Rozhdestvensky, K., Ryzhov, V., Semenov, N., Chemodanov, M.: Educational marine robotics in SMTU. In: Ronzhin, A., Rigoll, G., Meshcheryakov, R. (eds.) ICR 2016. LNCS, vol. 9812, pp. 79–88. Springer, Cham (2016). doi:10.1007/978-3-319-43955-6_11
8. Wooldridge, M.: An Introduction to MultiAgent Systems. Wiley, Hoboken (2002)
9. Keil, D., Goldin, D.: Indirect interaction in environments for multi-agent systems. In: Weyns, D., Van Dyke Parunak, H., Michel, F. (eds.) E4MAS 2005. LNCS, vol. 3830, pp. 68–87. Springer, Heidelberg (2006). doi:10.1007/11678809_5
10. Sun, R.: Cognition and Multi-Agent Interaction. Cambridge University Press, Cambridge (2006)

Author Index

Printed in the United States
By Bookmasters

Printed in the United States
By Bookmasters